Lecture Notes in Mathematics

Edited by A. Dold and E

1366

Norman Levitt

Grassmannians and Gauss Maps in Piecewise-linear Topology

Springer-Verlag
Berlin Heidelberg New York London Paris Tokyo

Author

Norman Levitt
Department of Mathematics
Rutgers, The State University
New Brunswick, NJ 08903, USA

Mathematics Subject Classification (1980): 57Q35, 57Q50, 57Q91, 57R20

ISBN 3-540-50756-6 Springer-Verlag Berlin Heidelberg New York
ISBN 0-387-50756-6 Springer-Verlag New York Berlin Heidelberg

Printed in Germany

Printing and binding: Druckhaus Beltz, Hemsbach/Bergstr.
2146/3140-543210

To my Parents

CONTENTS

0. Introduction

This monograph brings together a number of results centered on an attempt to import into the study of PL manifolds some geometric ideas which take their inspiration from the origins of differential topology and differential geometry, ideas from which many important aspects of fiber-bundle theory have developed. The reader is presumed to be familiar with the central role that the theory of fiber bundles has played in the study of differentiable manifolds for the past four decades. The central theme here has been that a wide class of geometric problems can be reformulated as bundle-theoretic problems. Typical results flowing from this approach have been the Cairns-Hirsch Smoothing Theorem; The Hirsch Immersion Theorem, together with its generalization, the Gromov-Phillips Theorem, and much of the important work in foliation theory. The great advantage of a reduction to bundle theory as has been generally been thought, is that the geometric problem has become a homotopy - theoretic problem from whence, with a little luck, it can be made into an algebraic problem.

It is also presupposed that the reader is conversant with the generalizations of classical vector bundle theory, generalizations which appropriate much of the machinery developed for differentiable topology for use in the study of PL manifolds, topological manifolds, homology manifolds, Poincare-duality spaces and so forth. In particular, the notions of PL bundle, PL block-bundle, topological bundle, spherical fibration (together with their stable versions) are assumed to be familiar territory. So, too, the classifying spaces (and canonical bundles) associated with these notions:
$BO(k)$ for vector bundles, $B\widetilde{PL}(k)$ for k-dimensional PL-bundles, $BPL(k)$ for PL k-block-bundles, $BG(k)$ for $(k-1)$ - spherical fibrations, and so forth.

I now wish to observe that these generalizations and the theorems that have exploited them have had a certain flavor, displaying, so to speak, an inclination to move into the homotopy theory as quickly as possible from the point of view of underlying constructions as well as that of ultimate results. A brief historical overview might make this clearer.

The notion of bundle and its applicability to topological questions goes back, of course, to Gauss, whose great work on curvature and its relation to the topology of surfaces exploits the Gauss map in its original and most literal sense. This of course is the map which, for any oriented surface immersed in 3-space, takes each point to the correctly-oriented unit normal vector to the surface at that point, the target space of the map being thought of as the standard unit 2-sphere.

In this century, the foundational work of Steenrod, Whitney, et. al. led to the formal definition of fiber bundles, with vector bundles along with principal Lie group bundles serving as the prime example. The discovery of the role of the Grassmann manifold as the "classifying space" for vector bundles preserved much of the original insight of Gauss' construction. As beginners in the subject soon learn, it helps one's intuition to picture vector bundles as tangent bundles to manifolds, particularly manifolds embedded or immersed in Euclidean space. In that case, one easily goes on to picture the classifying map, (frequently and quite appropriately called the "Gauss map") as that map which takes each point in the given n-manifold to the point in the appropriate Grassmannian corresponding to the unique n-dimensional linear subspace (of the given Euclidean space) parallel to the tangent space at the point.

In the intervening decades, generalizations of the notion of vector bundle have proliferated, and the notion of "universal classifying space" has become a familiar one for many contravariant homotopy functions beyond vector bundles and principal bundles. The chief tool here is E. Brown's Representability Theorem [Bro] and some of its generalizations, which guarantee that a homotopy functor is "representible" (i.e, has a classifying space within the category of CW complexes) under very unrestrictive conditions. In particular, Brown's Theorem is usually cited as the justification for asserting the existence of BPL, B Top, BG et.al.

Despite the beauty and usefulness of the Representability Theorem, however, I wish to assert that there is something problematical about its use in connection with intrinsically geometric problems. First of all, one sees that the classifing space B_F obtained for a given functor F is truly a "homotopy theoretic" object; it has no "natural" geometric structure and, indeed, is a geometric object only in the most shadowy and abstract sense. The same may be said of the map $X \to B_F$ classifying an element of F(X). This is no map at all strictly speaking, but rather a homotopy class of maps. In some sense, to the degree that we rely on the Representability Theorem, we "know" B_F or $[X,B_F]$ precisely as well as we know F or F(X). The roll of B_F as a space or an element of $[X, B_F]$ as a map is largely mataphorical. Note how far this is in spirit from the original Gauss construction, in which a specific geometric object (an embedded manifold) was seen to acquire an equally specific map into a concrete geometric object (the standard sphere), a map whose local properties, moreover, were of intense geometric interest. Gauss, after all, was not interested in the abstract classification of normal bundles of surface but rather in understanding the local geometry of curvature in its relation to global invariants.

The present work is a first attempt at recovering something of this spirit for the study of combinatorial manifolds. Combinatorical manifolds, after all, are by definition, objects which support specific geometric structures, namely triangulations (more specifically, metric triangulations where each simplex has a metric consistent with its convex linear structure). There is a rough but useful analogy: triangulated manifolds are to combinatorial manifolds as Riemannian manifolds are to differentiable manifolds. That comparison suggests, among other implications, that the local properties of a triangulation ought to bear some relation to the global invariants of the manifold.

The problem, of course, is to give this insight some concrete point. The view taken in these notes is that the local geometry of a triangulated manifold gives rise to a map (and the emphasis here is on map rather than homotopy class of maps) into a universal example which, so to speak, is constructed from all possible prototypes of local geometrices. In view of tradition and of the naturality of the construction we call this map a Gauss map.

This usage is further justified by the fact that the Gauss map, as we define it, carries the appropriate bundle information. That is, the Gauss map is naturally covered by a bundle map (in the appropriate category) of the tangent bundle of the manifold to some canonical bundle over the universal space (which is thus naturally to be thought of as a kind of "Grassmannian"). Carrying this analogical mode of thinking yet further, we might consider a triangulated manifold embedded in Euclidean space so that the embedding is a convex-linear map on each simplex. The analogy here is to smooth a submanifold of Euclidean space. One ought to suspect that, just as there is a natural Grassmannian which receives the Gauss map of the embedded smooth manifold, there might be a natural space which receives the equally natural Gauss map of the embedded manifold. This suspicion is quite justified. Again, prototypes of local geometries (where now the embedding in Euclidean space is to be taken into account) can be assembled to form the appropriate PL Grassmannian which in turn supports an appropriate canonical bundle.

Once embarked upon this mode of thinking, we find ourselves naturally drawn into generalizations and extensions of the main idea of constructing Grassmannians and Gauss maps to handle different kinds of underlying geometric situations. To name but one example by way of suggesting the flavor of our approach, we might consider whether a combinatorical manifold M admits a "bundle of Grassmannians" so that given an immersion V M, there will be a Gauss map from V to that "bundle" covering the immersion.

Leaving aside for the moment an exact enumeration of those geometrical considerations which give rise to "Grassmannians" and "Gauss maps", we come to the further problem of justifying such constrictions beyond the limited appeal of abstract ingenuity.

First of all, we shall exploit the notion that a Gauss map (in contradistinction to a homotopy-theoretic classifying map into a homotopy-theoretic classifying space) is both concrete and locally determined. This can be used to convert global information into local information, at least in principle. The analogy to be borne in mind here is to the Chern-Weil theorem [Mi-St] on characteristic classes of Riemannian manifolds. Just as a universal differential form in the classical Grassmannian pulls back (given a classical Gauss map) to a de Rham co-cycle representing a characteristic class, a "universal co-cycle" in one of our "PL" Grassmannians performs a similar function. (In the subsequent chapter-by-chapter outline, we shall address this point more specifically.)

Beyond this, we are interested in the relation between "geometrical structure" on manifolds and Gauss maps. Geometrical structure, in our sense typically means immersion of the manifolds into a given ambient space, possibly with additional conditions as to the "local geometry" of the immersion. In the smooth case, such geometric questions usually are phrased in terms of infinitesimal data, so that a "geometry" for the manifold may be most usefully thought of as a cross section of some bundle of map germs satisfying, say, some further condition defined in terms of a jet bundle to which the original germ-bundle maps via differentials. The simplest example is an immersion, which is of course a smooth map whose 1-jet is of maximal rank every- where. The thematic result here is the theorem of Hirsch, Gromov and Phillips, [P] which assures us in a large number of cases that a section of the jet bundle with the appropriate properties is sufficient evidence for the existence of a section of the germ bundle itself, whose differential has the same properties as the original section. Again taking the simplest example, Hirsch's original result [Hi] tells us that a map between manifolds is homotopic to an immersion if it can be covered by a map of tangent bundles of maximal rank everywhere (with some additional assumptions necessary in codimension 0).

Of course it is well known that the Hirsch Theorem admits a generalization into the PL category, with conditions being phrased in terms of PL tangent bundles. Yet if we wish to study immersions satisfying certain further restrictions, natural from the point of view of PL geometry, the general ideas of the Gromov-Phillips Theorem seem inadequate. There are no differentials, jet bundles etc. in the PL category.

However, we shall see that certain kinds of geometries on manifolds - certain kinds of immersions meeting local specifications - do correspond in natural ways to the Grassmannians we shall construct and, more particularly, to subspaces thereof. That is, an immersion whose local properties satisfy some restriction has a Gauss map whose image lies in an appropriate subspace of the Grassmannian. Thus, in the spirit of the Gromov-Phillips theorem but with much different constructions in hand, we may ask the converse question: Given an abstract map of a manifold to the indicated subspace of the Grassmannian, covered by a map from the tangent bundle to the cannonical bundle, can we then obtain an immersion with the appropropriate geometry? We shall prove theorems of this kind usually with the proviso that the manifold in question be open.

We shall also address further questions in a related vein having to do with smoothing theory and with piecewise-differentiable, rather than piecewise linear maps. We shall also consider versions of these results in the context of actions by finite groups. The reader may find the following outline useful.

Chapter 1. Local formulas for characteristic classes.
The main topic in this section is an exposition of the author's joint work with C. Rourke [Le-R] proving the existence of local formulas for rational characteristic classes of PL manifolds. The methodology here is thematic. A semi-simplical complex $|Q_n|$ is constructed which is the natural target of a Gauss map from triangulated n-manifolds with a local ordering of vertices. $|Q_n|$ naturally supports a canonical n-block bundle which receives a natural n- block bundle map from the tangent block-bundle of such a manifold, which map covers the Gauss map. The existence of characteristic classes for the canonical block-bundle easily leads to the existence theorem. The chapter also contains a generalization to homology manifolds as well as a brief discussion of various attempts to find a concrete formula for the Pontrgagin classes and L-classes.

Chapter 2. Formal links and the PL Grassmannian $\mathcal{G}_{n,k}$.
In this chapter we construct the "PL Grassmannian" $\mathcal{G}_{n,k}$, together with its canonical PL n-bundle $\gamma_{n,k}$. This is the natural Grassmannian for simplex-wise linear immersions of triangulated n-manifolds into R^{n+k}. It is shown how a Gauss map arises naturally and automatically for such immersions.

Chapter 3. Some variations of the $\mathcal{G}_{n,k}$ construction.
This chapter briefly explores the construction of spaces akin to $\mathcal{G}_{n,k}$ and appropriate to geometric situations other than simplex-wise linear immersions of triangulated manifolds. In particular maps more general than immersions and complexes more general than combinatorical

manifolds correspond to certain spaces defined similarly to $\mathcal{G}_{n,k}$.

Chapter 4. The immersion theorem for subcomplexes of $\mathcal{G}_{n,k}$.
In this section we define the notion of geometric subcomplex of $\mathcal{G}_{n,k}$. In spirit, this means a subcomplex which receives the Gauss map of manifolds immersed in such a way that additional geometric restrictions are observed. If $\mathcal{H}$ is such a subcomplex, we consider manifolds M^n whose tangent bundles map to the restriction to $\mathcal{H}$ of the canonical bundle $\gamma_{n,k}$. The main result, generalizations of which occupy much of the remaining text, is that if such a manifold be non-closed, then it will immerse in R^{n+k} so that the Gauss map has image in $\mathcal{H}$.

Chapter 5. Immersions equivariant with respect to orthogonal actions on R^{n+k}.
Here we generalize the result of the last chapter to deal with triangulated manifolds on which a finite group acts simplicially and with orthogonal actions by that group on R^{n+k}. (The group then automatically acts on $\mathcal{G}_{n,k}$ as well.) The idea is to obtain equivariant immersions subject to additional geometric conditions corresponding to an invariant geometric subcomplex $\mathcal{H}$. The result holds for manifolds satisfying the so - called Bierstone condition.

Chapter 6. Immersions into triangulated manifolds.
This chapter contains the thesis work of my student Regina Mladineo. As the title suggests, we study immersion theory where the target space is now a triangulated manifold rather than Euclidean space. We start by constructing, for a triangulated manifold, an analog to the Grassmannian bundle associated to the tangent bundle of a smooth manifold. If W^{n+k} is triangulated we construct $\mathcal{G}_{n,k}(W)$ which is the natural target of a Gauss map from M^n, where M^n is a triangulated manifold immersing in W^{n+k} in general position with respect to the triangulation. Here it is also assumed that inverse images of simplices of W are subcomplexes of M and that the map

is simplex-wise convex-linear. In point of fact, $\mathcal{G}_{n,k}(W)$ is not a fiber bundle over W but rather a semisimplicial complex assembled from a collection of copies of $\mathcal{G}_{n-r,k}$ with one copy for each simplex of W of codimension r. Geometric subcomplexes are then defined and it is shown that a result analogous to that of Chapter 4 can be obtained. If W and M are further equipped with simplicial actions by a finite group then the analog to the result of Chapter 5 can be obtained as well.

Chapter 7. The Grassmannian for piecewise-smooth immersions. Here we broaden our considerations to study PL manifolds equipped not with a triangulation but rather with a stratification which is "linkwise simplicial" and where each stratum is provided with a smoothness structure so that inclusions of strata into higher strata are smooth. If we consider piecewise-smooth immersions of such manifolds M into Euclidean space R^{n+k}, it is natural to look for an appropriate notion of Grassmannian. This space, which we designate $\mathcal{G}^c_{n,k}$ turns out to be closely related to the $G_{n,k}$ of previous chapters. In fact, $\mathcal{G}^c_{n,k}$ is $\mathcal{G}_{n,k}$ retopologized as a the geometric realization of a simplicial space rather than a simplicial set. A theorem analogous to the main result of Chapter 4 is obtained.

Chapter 8. Some applications to smoothing theory. This chapter represents a detour from the main thrust of the foregoing Chapters 2-7 in that we are no longer concerned with immersion theory but with smoothing theory. We begin with the construction of a space A^{ord} which is, in some sense a simpler version of the $|Q_n|$ of chapter 1 and of $\mathcal{G}_{n,k}$ as well. A^{ord} is the natural target of Gauss map from a locally ordered triangulated manifold M^n, yet, N.B., it is not constructed with a view to supporting a canonical PL bundle. A^{ord} has, so to speak, one i-cell for each possible ordered triangulation of S^{i-1}. We then go on to construct another space A^{Br} which has one i-cell for each "Brouwer structure" on the cone on an ordered, triangulated S^{i-1}, where a Brouwer structure means a simplex-wise linear embedding in R^i. A^{Br} is then retopologized (as $\mathcal{G}_{n,k}$ was retopologized to produce $\mathcal{G}^c_{n,k}$) to yield yet another space A^{CBr}. A^{CBr} maps naturally into A^{ord}. Our theorem is that M^n is smoothable if and only if there is a homotopy lift in the diagram

$$\begin{array}{ccc} & & A^{CBr} \\ & & \downarrow \\ M^n & \rightarrow & A^{ord}. \end{array}$$

What is interesting about this result is that the property sought has no a priori connection with bundle theory.

Chapter 9. Equivariant piecewise differentiable immersions. We resume the main theme of these notes by considering piecewise-smooth manifolds supporting a compatible finite group actions and equivariant immersions into a Euclidean space on which the group acts orthogonally. We generalize the result of Chapter 7 just as Chapter 5 generalized that of Chapter 4.

Chapter 10. Piecewise differentaiable immersions into Riemannian manifolds.

We now consider piecewise-smooth immersions where the target is a smooth manifold equipped with a Riemannian metric. For such spaces W^{n+k} we construct an "associated Grassmannian bundle $\mathcal{G}^c_{n,k}(W)$ (now truly a bundle) whose fiber is the $\mathcal{G}^c_{n,k}$ of chapter 7. $\mathcal{G}^c_{n,k}(W)$ is the natural target of a Gauss map from M^n when M^n is piecewise-smoothly immersed. As Chapter 6 generalized the results of Chapters 4 and 5, this Chapter generalizes Chapters 7 and 9.

A brief glossary of important definitions and constructions is provided in the appendix.

1. Local Formulae for Characteristic Classes

The point of view which looks at the characteristic classes of a manifold as global summaries of local data is a rather old one. Insofar as Stieffel-Whitney classes are concerned, this approach may be said to have been born with the subject. In particular, though Stieffel-Whitney classes were devised in connection with vector bundles and smooth manifolds, it became clear early on that the definition readily extended to combinatorial manifolds. [In fact, via the definition $w_i = (\cup \Phi)^{-1}Sq^i\Phi$, Φ the $Z/2Z$ Thom class of the bundle in question, it is easily seen that Poincare duality spaces have well-defined Stieffel-Whitney classes as well]. But the more interesting aspect of the definition of w_i on a combinatorial manifold, or, more correctly, a combinatorially triangulated manifold, is that the definition is local. We remind the reader how the formula works.

Let M^n be a combinatorially triangulated n-manifold. With respect to this fixed triangulation T, we have the first barycentric subdivision T'. The formula for w_i may be viewed as giving an i-co-cycle (co-efficients in $Z/2Z$) for the cell structure on M^n dual to T'. Alternatively, we may read the formula as giving a representative for the n-i homology class w^*_{n-i} Poincare dual to w_i, by specifying an (n-i) cycle in T' itself. The formula is extraordinarily simple:

Let $\gamma^*_{n-i} = \sum \tau_{n-i}$ where τ_{n-i} ranges over all the (n-i)-simplices of T' (in int M if $\partial\partial \neq 0$). Then

<u>1.1 Theorem</u> (Whitney [Whn]; see also [Ch1], [H-T]). γ^*_{n-i} is a $Z/2Z$ cycle whose homology class is w^*_{n-i} $H_{n-i}(M, \partial M; Z/2Z)$.

Thus, one may read off directly, on the chain level, the Poincare duals of the standard Stieffel-Whitney classes. If one wishes to translate this into a corresponding statement about co-cycle representatives (in the dual cell structure) for the Stieffel Whitney cohomology classes themselves, it is useful to order the triangulation, at least so that each simplex is linearly ordered. The ordering ℓ canonically defines a subdivision map $\lambda: T' \to T$, and so we obtain a cycle $\lambda_* \gamma^*_{n-i} \in C_*(T, T \cap \partial M; Z/2Z)$. If we let γ^i be defined (with respect to the cell structure T^* Poincare dual to T) by $\gamma^i(\sigma^*)$ = (number of (n-i)-simplices in $\lambda^{-1}\sigma \subset T'$) (mod 2). Then

<u>1.2 Corollary</u>. γ^i is a cocycle in $C^i(T^*, Z/2Z)$ representing the Stieffel-Whitney class $w^i(M)$.

Note that the value of γ^i on a dual i-cell σ^* depends only on the structure of the ordered simplicial complex $st(\sigma)$. This sets the pattern for our generalization, at least on the level of existence theorems, to arbitrary characteristic co-homology classes of PL n-manifolds.

Let M^n be a PL manifold with a combinatorial triangulation T.

<u>1.3 Definition</u>. A <u>local</u> <u>ordering</u> for T is a partial ordering of the vertices of T such that each star $st(\sigma^k, T)$ (abbreviated $st(\sigma)$) (σ^k a k-simplex of T) is thereby linearly ordered.

Abstractly, an <u>n-star</u> of codimension i, $i \leq n$, shall mean a complex of the form $\Delta^{n-i} * \Sigma^{i-1}$, where Δ^{n-i} is the standard n-i simplex and Σ^{i-1} denotes a combinatorially triangulated (i-1)-sphere (= 0 if i = 0). An <u>ordered</u> codimension-i n-star is such an object with a linear ordering of its vertices, and an <u>oriented</u> star means one where Σ^{i-1} has been given an orientation ω. <u>Isomorphism</u> of ordered stars means a simplicial isomorphism preserving both the ordering and the factors Δ, Σ of the join.

1.4 Definition. A local-ordered formula for an i-dimensional co-chain with coefficients in G is a function ϕ defined on (isomorphism classes of) oriented, ordered, codimension-i n-stars taking values in G. It is further stipulated that
$\phi(\Delta^{n-i} * \Sigma^{i-1}, \omega) = - \phi(\Delta^{n-i} * \Sigma^{i-1}, -\omega)$.

A local formula is merely a local-ordered formula such that $\phi(\Delta^{n-i} * \Sigma^{i-1}, \omega)$ depends only on the simplicial structure of Σ^{i-1} and not on the ordering of $\Delta^{n-i} * \Sigma^{i-4}$.

If M^n is a manifold with a locally ordered triangulation T, it is clear that for any (n-i)-simplex σ^{n-i}, the star $st(\sigma,M)$ may be regarded as an ordered, co-dimension-i n-star $\sigma^{n-i} * lk(\sigma,M)$. Thus, given a local ordered formula ϕ for an i-dimensional G-cochain, we obtain a co-chain $\phi(T) \in C^i(T^*;G)$, where T^* denotes the call structure on M^n Poincare' dual to T. We note that for $\partial M \neq \emptyset$, T^* is a cell structure on a deformation retract of M rather than the whole of M, but this is a minor point. We see this by noting that for each dual i-cell σ^* (σ an n-i-simplex of T, interior to M should $\partial M \neq \emptyset$) an orientation o may be regarded as an orientation ω of $lk(\sigma,M) = lk(\sigma)$. Thus the assignment σ^*, $o \to \phi(\sigma^* lk(\sigma), \omega) \in G$ defines an i-co-chain in the (oriented) co-chain theory $C^*(T^*;G)$. We denote this class $\phi(T)$.

We consider an i-dimensional characteristic class c for n-dimensional PL manifolds, i.e. an element $c \in H^i(B\widetilde{PL}(n); G)$.

1.5 Definition. The local (ordered) formula ϕ is said to represent c if and only if, for all combinatorial manifolds M^n, and all triangulations T, the co-chain $\phi^i(T)$ is a co-cycle with $[\phi(T)] = c(M) \in H^i(M,G)$.

Our main result is:

1.6 Theorem. [Le-R] Given any characteristic class $c \in H^i(B\widetilde{PL}(n),G)$, there exists a local-ordered formula ϕ representing c.

A special case of particular interest occurs when G is the rational numbers. We then have the following relevant corollary.

<u>1.7 Corollary</u>. Let G be a divisible group, $c \in H^i(B\widetilde{PL}(n);G)$. Then there is a local formula representing c.

Proof: By 1.6, let ϕ_1 be a local-ordered formula representing c. Define the local (unordered) formula ϕ on an (unordered) co-dimension i n-star $\Delta^{n-i} * \Sigma^{i-1}$ by

$$\phi(\Delta^{n-i} * \Sigma^{i-1}) = \frac{1}{q!} \sum_{\Lambda} \phi_1(\Delta^{n-i} * \Sigma^{i-1}, \Lambda)$$

where q is the number of vertices of $\Delta^{n-i} * \Sigma^{i-1}$ and Λ ranges over all possible linear orderings of these vertices. Then, for a manifold M with (finite) triangulation T, and local ordering π, we have $\phi_1(T,\pi)$ a co-cycle representing $c(M)$. Thus, if m = number of distinct local-orderings of T, $\psi(T) = \frac{1}{m} \sum_{\pi} \phi_1(T,\pi)$ is a co-cycle also representing $c(M)$. Clearly, given a simplex $\sigma^{n-k} \subset \operatorname{int} M$ and an orientation o on σ^*, we have $\psi(T)(\sigma^*, o) = \phi(\sigma^* lk(\sigma), \omega)$ (where o corresponds to W).

<u>1.8 Corollary</u>. There is a local formula representing any rational characteristic class for PL manifolds; in particular the rational Pontrjegin class p_i and the rational L-class L_i are so represented.

<u>1.9 Corollary</u>. Suppose M^n is triangulated by T so that for any $(n-i)$-simplex $\sigma \subset \operatorname{int} M$, $\ell k(\sigma)$ admits an orientation-reversing simplicial self-homeomorphism. Then all i-dimensional rational characteristic classes of M must vanish.

Proof: Let ϕ be a local formula with co-efficients in Q and $\Delta^{n-i} * \Sigma^{i-1}$ be a codimension-i n-star with Σ^{i-1} admitting an orientation reversing simplicial self-homeomorphism. Then given an orientation ω on Σ^{i-1}, it follows that $\Delta^{n-i} * \Sigma^{i-1}, \omega$ is isomorphic to $\Delta^{n-i} * \Sigma^{i-1}, -\omega$. So $\phi(\Delta^{n-i} * \Sigma^{i-1}, \omega) = \phi(\Delta^{n-i} * \Sigma^{i-1}, -\omega)$. But $\phi(\Delta^{n-i} * \Sigma^{i-1}, \omega) = -\phi(\Delta^{n-i} * \Sigma^{i-1}, -\omega)$. So $\phi(\Delta^{n-i} * \Sigma^{i-1}, \omega) = 0$.

Thus, $\phi(T) \equiv 0$, with the given hypothesis on T. Therefore, since ϕ may be chosen to represent any given rational characteristic class, all such classes vanish on M.

Before moving to the actual proof of 1.6, some philosophical discussion is in order. Obviously, the primary inspiration for conjecturing that local (ordered) formulae must exist is the example cited in 1.1 - 1.2 above. Furthermore, the temptation to generalize to, say, rational Pontrjagin classes or L-classes is further strengthened by the differential-geometric result that for smooth manifolds provided with Riemannian structures (or merely affine connections) these classes (with real co-efficients) are canonically represented in de Rham cohomology by "locally determined" differential forms. That is, given a Riemannian manifold M^n, the real Pontrjagin class $p_i(M)$ is represented by the Chern-Weil form $P_i(M) \in \Omega^i(M)$, $dP_i = 0$. Moreover P_i is "local" in the sense that for any open set U of M, $P_i(U) = P_i(M)|U$. For details the reader may consult the book of Milnor and Stasheff [M-S].

It will be a continuing theme of this monograph that the assignment of a specific triangulation to a PL manifold is in many ways analogous to choosing a specific Riemannian metric for a smooth manifold. That is, one is endowing a "topological" object with a specific, rather rigid geometry. The thematic principle then emerges that global information about the manifold should then be seen as a summary, so to speak, of local contributions determined by local geometry. Both the Chern-Weil forms for real characteristic classes and the Whitney cycle formula for (dual) Stieffel Shitney classes may be seen as illustrative examples. The conjecture that arbitrary PL characteristic classes are represented by local formulae therefore becomes quite natural. In particular, one expects that the local infinitesimal data on a Riemannian manifold giving rise to the Chern-Weil forms ought to be replaced by "singular" data, i.e. discrete contributions for each bit of relevant local geometry. For an i-dimensional class the "relevant" bits should be the local geometry in the neighborhood of each n-i simplex, in other words, the simplicial structure of the links of such simplices. The example of the Stieffel-Whitney classes suggests, at least, that ordering data should figure as well for such a local formula, at least in the absence of an averaging procedure like that in Cor. 1.7.

Historically, the first example of such a local formula comes from the papers of Gabrielov, Gelfand and Lossik [GGL] on the determination of a cocycle representing p_1 of a smoothly triangulated smooth manifold which turns out to depend merely on the local combinatorial structure of the triangulation. The procedure is complicated and somewhat obscure, although clarified somewhat by the papers of MacPherson [Mac] and D. Stone [St_1, St_2]. We shall not describe this construction here, although, at the end of this section, we shall make some remarks on Gabrielov's attempt to extend these methods to higher Pontrjagin classes. It is noteworthy, however, that, although p_1 is an integral class on PL manifolds, the method of [GGL] do not seem to result in a local-ordered formula for an integral representing cocycle.*

Cheeger [Ch_2] has attacked, with some success, the problem of finding local formulae for the real L-classes, and we shall also briefly describe the general idea of his approach after proving 1.6.

We must take note, at this point, that Theorem 1.6 is purely an existence theorem, as the proof will make clear. No attempt is made to describe an explicit construction of the local (ordered) formula for a given class. Nevertheless, the existence proof is surprisingly quick and elegant, and demonstrates the power of the viewpoint taken

*See [Mi] for a computational example. See [Le_2] for a somewhat different approach.

in this monograph as a whole: Given some notion of explicit geometry on a manifold, it becomes possible to replace the idea of "classifying space" for the appropriate kind of bundle by a "Grassmannian." That is "classifying spaces" are, traditionally, objects in the homotopy category whereas a "Grassmannian" means a specific space with its own explicit geometry. At the same time, the "classifying map" for the tangent bundle of a manifold (defined up to homotopy) is refined, in the presence of geometry, into a "Gauss map" i.e. a specific, canonical map into the Grassmannian which somehow keeps track of the local geometry of the manifold.

The proof of 1.6, which we now give, recapitulates that to be found in the paper of the author an C. Rourke [L-R].

First, some terminology.

An s-ball is a linearly-ordered simplicial complex K such that $|K|$ is a Euclidean ball.

An <u>s-cell complex</u> is a partially ordered simplicial complex K, together with a family of subcomplexes $\{L_i\}$ such that

(1) Each L_i is totally ordered and, as well, an s-ball (of some dimension).

(2) $|K|$, $\{|L_i|\}$ is a cell complex (as in [R-S, p. 3].

Thus, an s-cell complex is basically a cell complex, further triangulated so that each cell is a subcomplex, and with the triangulation partially ordered so that the subcomplex for any cell is linearly ordered.

An isomorphism $h: K_1 \to K_2$ between s-cell complexes is a simplicial isomorphism preserving the ordering on each cell.

As an example, consider a combinatorially triangulated manifold with a local ordering. If K is the triangulation, the Poincare' dual cell structure K^* then becomes an s-cell complex, in the sense that K' has a derived local ordering (viz., the lexicographic ordering on its vertices arising from the ordering on K); moreover, each dual cell of K^* is a subcomplex of K' linearly ordered by the partial order of K', and is therefore an s-ball.

An s-block bundle ξ over an s-cellcomplex K (of fiber dimension k) consists of the following:

(1) A partially-ordered complex Q with $K \subset Q$, (preserving ordering)

(2) Linearly-ordered subcomplexes $\{R_i\}$, (one for each cell $L_i \subset K$) with $L_i \subset R_i$ and R_i an s-ball such that: $|Q|,\{|R_i|\}$ forms a k-block bundle over $|K|,\{|L_i|\}$.

An isomorphism of s-block bundles is a simplicial isomorphism which is also a block-bundle isomorphism and which preserves the linear order on each block.

To take an example, let $M^n \subset W^{n+k}$ be a (locally flat) embedding of PL manifolds. Suppose P is a locally-ordered triangulation of W such that M^n is triangulated by the subcomplex K. Recall the construction of [R-S] of the normal block bundle $\nu_W(M)$ over M of the embedding: If σ is a simplex of K, and σ^*, σ^* represents the dual cell in K^*, P^* respectively, then $E = \bar{\sigma}^*$ is a k-block-bundle over $M = |K| = |K^*|$ with blocks $\bar{\sigma}^*$ over σ^*. As we have seen, K' is an s-cell complex with cells σ^*, and, since the cells $\bar{\sigma}^*$ become s-balls (when regarded as subcomplexes of P') $\nu_W(M)$ acquires the structure of an s-block-bundle over K^* (of fiber dimension k).

From this example, it becomes possible to see how, given a locally ordered triangulation K of M, the tangent block-bundle $\tilde{T}M$ (so-called to distinguish it, as a formality, from the tangent PL bundle TM) acquires the structure of an s-block bundle. For given the ordering on K, we may triangulate $|K|\times|K| = M\times M$ in a specific way (i.e. the cell $\sigma\times\tau$ becomes triangulated in a standard way, for any pair σ,τ of simplices, since σ,τ are linearly ordered). Moreover, $K\times K$ is locally-ordered by the induced lexicographic order on its vertices. Further, the diagonal map $M \xrightarrow{\Delta} M\times M$ is a simplicial map $K \xrightarrow{\Delta} K\times K$, and thus $\nu_{M\times M}(\Delta M)$ acquires an s-block bundle structure as above. But, of course $\nu_{M\times M}(\Delta M)$ is, by definition, $\widetilde{T}M$.

Our purpose now is to construct a canonical s-block bundle γ_n over a universal space Q_n which will classify s-block bundles of fiber dimension n.

To this end, some further terminology:

An s-Cell is an s-cell complex with a single top-dimensional cell of which all others are faces.

The category $\mathcal{S}$-Cell is defined by taking, as objects, isomorphism classes of s-Cells and, as morphisms, face inclusions.

An $\mathcal{S}$-Cell set is a contravariant functor from $\mathcal{S}$-Cell to the usual category C of sets.

[The reader may think of the notion of $\mathcal{S}$-Cell set as a variant of the familiar notion of Δ-set (i.e., semi-simplicial set.)]

Given an $\mathcal{S}$-cell set Q, we define the geometric realization $|Q|$ of Q as follows:

Given an isomorphism class A of s-Cells, consider $|A| \times Q(A) = \Gamma(A)$, where $Q(A)$ is a set with the discrete topology form the union $\bigcup_{A\in Ob(\mathcal{S}\text{-Cell})} \Gamma(A)$ and then take the identification space coming from identifying $|B| \times \{Q(f)(x)\}$ with $|f|(|B|) \times \{x\}$ $|A| \times \{x\}$ whenever $B \xrightarrow{f} A$ is face-morphism of s-Cells and x $Q(A)$. This defines $|Q|$.

[The reader may compare this with the standard procedure for forming the geometric realization of a Δ-set.]

Notice that the standard model category for Δ-sets is a full subcategory of $\mathcal{S}$-Cell. I.e. if Δ has as objects the standard ordered simplices of each dimension and, as morphisms, order preserving face inclusions, then $\Delta \subset \mathcal{S}$-Cell as a full subcategory.

There is a natural functor from Δ-sets to $\mathcal{S}$-Cell sets; given the Δ-set Q, extend it to an $\mathcal{S}$-Cell set by $Q(A) = \emptyset$ for $A \notin Ob(\Delta)$.

On the other hand, there is a functor from $\mathcal{S}$-Cell sets to Δ-sets since, given the $\mathcal{S}$-Cell set Q, $|Q|$ is the union of s-cells. Thus it is the union of (ordered) simplices and thus, in an essentially unique way $|Q| = |Q_\Delta|$ for some Δ-set Q_Δ. Thus, if Q initially arises from a Δ-set, we have $Q_\Delta = Q$.

Let $CW_{\mathcal{S}}$ denote the category whose objects are $\mathcal{S}$-Cell sets and whose morphisms are homotopy classes of maps of geometric realizations.

<u>1.10 Lemma</u>. $CW_{\mathcal{S}}$ is naturally equivalent to the category CW of CW-complexes and homotopy classes of maps.

Proof: We cite the well-known fact that CW is naturally equivalent to the category of Δ-sets and homotopy classes of maps of geometric realizations. By the facts cited above, there is a natural homeomorphism, $|Q| \cong |Q_\Delta|$ for any $\mathcal{S}$-Cell set Q, thus defining a natural map from $CW_{\mathcal{S}}$ to CW. By our standard way of regarding a Δ-set as an $\mathcal{S}$-Cell set, we get a map $CW \to CW_{\mathcal{S}}$. The composites $CW \to CW_{\mathcal{S}} \to CW$ and $CW_{\mathcal{S}} \to CW \to CW_{\mathcal{S}}$ obviously induce bijections on isomorphism classes of objects and morphisms.

We now define the universal s-block bundle γ_n over Q_n. Let $\mathcal{Q}_n$ be the $\mathcal{S}$-Cell set which assigns to each s-Cell A the set of isomorphism classes of s-block bundles over A of fiber dimension n. It is clear that restriction defines the requisite face maps to make $\mathcal{Q}_n$ a contra-variant functor, and thus an $\mathcal{S}$-cell set. Q_n denotes the obvious s-cell structure on $|\mathcal{Q}_n|$.

Let K be an s-cell complex and ξ an s-block bundle over K. We wish to define a canonical map (of -cell complexes). Canonically, $|K|$ is $|\mathcal{K}|$ for some $\mathcal{S}$-Cell set $\mathcal{K}$. Thus, for any $x \in \mathcal{K}(A)$ we have the assignment $x \to$ isomorphism class of $\xi|A \times \{x\}$ ($A \times \{x\}$ is naturally a sub-s-cell complex of K). Thus we have a transformation $\mathcal{K} \to \mathcal{Q}_n$ which we may think of as a map $i_\xi : K \to Q_n$.

Now γ_n is defined over Q_n by letting the block over $A \times \{y\}$ be y (up to isomorphism for any $y \in \mathcal{Q}_n(A)$, A any $\mathcal{S}$-cell. ($A \times \{y\}$ is a sub-s-cell complex of Q_n).

Clearly $i_\xi : K \to Q_n$ induces ξ from γ_n.

Forgetfully, γ_n over Q_n has the structure of an ordinary block-bundle, hence there is a homotopically unique classifying map $c: |Q_n| \to B\widetilde{PL}(n)$.

<u>1.11 Proposition</u>. c is a homotopy equivalence, i.e. $|Q_n| \simeq B\widetilde{PL}(n)$.

[Remark: 1.11 is not essential to the proof of 1.6, but is worth noting in passing.]

Proof: Let K be a simplicial complex. We must show that concordance classes of n-dimensional block bundles over K are in 1-1 correspondence with $[|K|, |Q_n|]$, which will suffice to prove the proposition.

Let ξ be a block bundle over K of fiber dimension n. Then we may triangulate the total space E so that the blocks over the original simplices of K are subcomplexes. Upon ordering this triangulation, we make ξ into an s-bundle ξ_a. Thus ξ_a is induced by a map $\iota_\alpha : K_a \to Q_n$ where K_a denotes a suitable s-cell complex whose cells are ordered subdivisions of the simplices of K. Moreover, given another such ordered triangulation of E, and resulting map $\iota_\beta : K_b \to Q_n$ we may triangulate $E \times I$ with suitable ordering, so as to agree with the given ordered triangulations on either end (making $|K| \times I$ into an s-cell complex L whose cells are subdivisions of the various $|\sigma| \times I$, σ a simplex of K.) Thus we get an s-block bundle η over L and, by the canonicity of classifying s-cell-complex maps for s-block-bundles, a map $\iota_\eta : L \to Q_n$ extending ι_α, ι_β. Thus $\iota_\xi = |\iota_\alpha| \quad |\iota_\beta| : |K| \to |Q_n|$ is well defined up to homotopy. I.e. we have a map $[|K|, BPL(n)] \to [|K|, |Qn|]$.

To show this is a bijection we need the following proposition. Let Q_n^{Δ} denote the s-cell-structure on Q_n coming from its simplicial structure.

<u>1.12 Lemma</u>. There is an s-cell structure on $|Q_n| \times I$ which is Q_n on one end and Q_n^{Δ} on the other.

Proof: We prove this by induction on the skeleta of Q_n. Our inductive hypothesis is that there is an s-cell structure on $|Q_n| \times \{0\} \cup |Q_n^{(k)}| \times I$ which agrees with the original s-cell structure on $|Q_n| \times \{0\}$ and with the simplicial subdivision of $Q_n^{(k)}$ on $|Q_n^{(k)}| \times \{1\}$. We further assume that the cells C not in either end are such that $|C| = |L| \times I$ for some cell L of Q_n. This obviously holds for $k = -1$. Assuming the hypothesis true, then, for a given k, we see that we may show it as well for $k+1$ as follows. For a $(k+1)$-dimensional cell L of Q_n, consider $|L| \times I$; we triangulate its boundary by taking the given triangulation of $|\dot{L}|\times I \cup |L|\times\{0\} \cup |L|\times\{1\}$. We triangulate $|L|\times I$ itself as c bdy($|L|\times I$). We order the triangulation of $|L|\times I$ consistent with the existing orderings, and then order the triangulation of $|Q_n| \times \{0\} \cup |Q_n^{(k+1)}| \times I$ consistently with all of these sub-orderings. Thus, as an s-cell, $L\times I$ has as k-cells of its boundary $L\times\{0\}$, $\sigma \times 1$, for σ a simplex of L, and suitable s-ball structures on $J\times I$ where J are the k-1 cells of L. This proves the lemma.

Let R denote the s-cell structure on $|Q_n| \times I$ guaranteed by the lemma. It follows, (by an argument similar to that which showed the existence of the map $[|K|, BPL(n)] \to [|K|, Q_n]$) that if we extend the underlying block bundle of γ_n to δ over R, and if we put any s-block structure on $\delta|Q_n^{\Delta}$ (with respect to the triangulation), we may extend the s-block structure to all of R. Let $r: R \to Q_n$ (extending id Q_n) classify this structure.

Now consider a map $f:|K| \to |Q_n|$. Simplicially approximate f by a map $g: K \to Q_n^{\Delta}$, and then note that $g^*(\delta|Q_n^{\Delta})$ acquires an s-block-bundle structure once some suitable order on a subdivision of K has been chosen. Moreover, $(r|Q_n^{\Delta})\circ g$ is the classifying map for this structure. But then $|(r|Q_n^{\Delta})\circ g|$ is homotopic to f as a map $|K| \to |Q_n|$. So we have an s-block bundle over some subdivision of K (i.e., the s-balls of the s-cell complex are the simplices of the subdivision) whose classifying map is f up to homotopy. We then may easily construct by amalgamation an s-block-bundle and over K itself (i.e., the s-balls are the simplices of K) whose classifying map is homotopic to f as well. This shows that $[|K|, BPL(n)] \to [|K|, |Q_n|]$ is onto. Since a map $|Q_n| \to BPL(n)$ exists classifying γ_n as a block bundle it follows that $[|K|, BPL(n)] \to [|K|, |Q_n|]$ must be injective, and the proposition is proved.

The proof of 1.6 itself, immediately below, is quite straightforward, given the construction of Q_n.

Given a locally-ordered triangulation T on the PL manifold M^n, we have seen how $\tilde{T}M$ acquires an s-block bundle structure (over the s-cell complex on M whose s-balls are the cells of T^*, triangulated and ordered as subcomplexes of T'.) Thus we get a well-defined s-cell-complex map $\tau:T^* \to Q_n$. (Again, we note the technical point that for $\partial M \neq \emptyset$, T^* is a cell structure on $M_0 = M-$ (collar on ∂M)). Note that τ is well defined as a cellular map, i.e., as a map $M \to |Q_n|$, it is specified pointwise, given the locally-ordered triangulation T, and not merely as a homotopy class. Thus we are justified in thinking of τ as a "Gauss map" (appropriate to the geometry of T), rather than as a "classifying map" which merely records tangent bundle information in the homotopy category.

Now let c be an i-dimensional characteristic class for PL n-manifolds, i.e. $c \in H^i(B\widetilde{PL}(n); G)$. Then $c(\gamma_n)$ is an element of $H^i(Q_n; G)$ and hence is represented by an i-dimensional cellular (oriented) co-chain $\Gamma \in C^i(Q_n; G)$. Thus $\tau^\#\Gamma \in H^i(T^*;G)$ represents $c(M^n)$. But note that τ as a global map, is the union of locally-defined maps on the cells of T^*. That is, over σ^*, the map $\tau|\sigma^* = \tau_\sigma$ is determined purely by the s-clock bundle $TM|\sigma^*$, and it is clear that, as an s-block bundle, this depends merely on the ordering of $st(\sigma,K)$. That is, given an ordered co-dimension j n-star $\Delta^{n-j} * \Sigma^{j-1}$, we may regard it as an ordered manifold D, and hence get a classifying map $d:D^* \to Q_n$, D^* = Poincare dual of $\Delta * \Sigma$. An orientation ω on Σ is tantamount to an orientation o on the j-cell Δ^* of D dual to Δ^{n-j}, hence, for $j = i$, we may define $\phi(\Delta^{n-i} * \Sigma^{i-1}, \omega)$ as $d^\#\Gamma(\Delta^*, 0)$. This locally-ordered formula obviously has the property

$$\phi(T) = \tau^\#\Gamma$$

and hence ϕ represents the characteristic class c. The proof of 1.6 is therefore complete.

<u>1.13 Remark</u>. As noted before, the proof of 1.6 is purely an existence theorem. Construction would amount to construction of a specific co-cycle representative on the "universal example" Q_n.

Before discussing other approaches to the problem of constructing local formulae, let us observe that 1.6 and its corollaries may be extended to a somewhat wider context. Heretofore, we have restricted our attention to PL manifolds and combinatorial triangulations thereof, but note that characteristic-class theory is of interest as well in studying homology manifolds (or rational homology manifolds).

Recall that an (integral) homology n-manifold is a simplicial complex such that $\ell k(\sigma^{n-i})$ has the integral homology of an $(i-1)$-sphere. Replacing the integers by some other unitary ring A of coefficients yields the definition of A homology manifolds (e.g., rational homology manifolds). The notion of (A)-homology-manifold with boundary may also readily be defined in the usual way.

Given an A-homology manifolds a dual "cell" structure exists, generalizing the Poincare' dual of a triangulation of a PL manifold. That is, given the A-homology manifold K, the cell σ^* is defined to be the star of b_σ in the first barycentric subdivision of $b_\sigma * \ell k(\sigma)$. Of course, σ^* is merely a cell in the "homology" sense since its boundary is a (A)-homology sphere.

There is a natural block bundle theory associated to A-homology manifolds. Omitting details, we point out that the "cells" in the base space are "cells" in the same sense that "dual cells" of an A-homology manifold are, and a similar condition holds for blocks over cells. It may then be seen that this kind of block-bundle theory has a classifying space [M-M]. We use $Bh_A(n)$ to denote the classifying space for A-homology block bundles of fiber dimension n. Further, it may also be seen that an A-homology manifold has a tangent bundle; the construction is essentially the same as for PL manifolds, i.e., one puts an A-homology block-bundle structure on a regular neighborhood of ΔK in $K \times K$. Therefore, an A-homology manifold M^n comes equipped, as it were, with a classifying map $TM: M \to Bh_A(n)$ and characteristic classes for such manifolds are thus defined in the usual way simply by setting,(for any cohomology class $c \in H^i(Bh_A(n);G))$, $c(M) = TM^*c \in H^i(M;G)$. We note, as an example of special interest, that for $A = G = Q$, the rational Pontrjagin classes p_i, or the rational L-class ℓ_i may be defined for rational homology manifolds since for their definition one only needs Poincare duality (with twisted rational co-efficients) and a suitable transversality theorem in the rational-homology-manifold category.

In any case the notion of local (ordered) formula for a given characteristic class is readily extensible to A-manifolds. We briefly sketch the procedure. First of all, Def. 1.3 (local ordering) may be applied unchanged to A-manifolds. Similarly we may alter the notion of codimension i n-star to mean a complex of the form $\Delta^{n-i} * \Sigma^{i-1}$ where Σ^{i-1} is now merely an A-homology manifold with the A-homology type of the $(i-1)$-sphere. Thus one may speak in this context of ordered n-stars, as well as oriented ones, keeping in mind that an orientation of $\Gamma^{n-i} * \Sigma^{i-1}$ now means an orientation of Σ^{i-1} for the co-efficient groups A.

Equally transparent is the adaptation of the notion of local (ordered) formula, viz., a local ordered formula for an i-co-chain, with coefficients in G is now taken to mean a function defined on isomorphism classes of ordered, oriented codimension-i n-stars, (in the newly modified sense) which takes values in the commutative group G and which respects change of orientation. A local formula, as before, is one which disregards ordering.

Local (ordered) formulae obviously specify G-valued i-co-chains on any (locally ordered) triangulated n-dimensional A-homology manifold. Here, of course, the co-chains are defined on a certain A-module chain complex not necessarily having the integral homology of M. Thus, it makes sense to speak of a local (ordered) formula ϕ as representing a characteristic class $c \in H^i(Bh_A(n);G)$. Furthermore, it is natural to try to extend the result of Theorem 1.6 to the larger context of A-homology manifolds.

<u>1.14 Corollary</u>. Given an i-dimensional characteristic class c for A-homology n-manifolds, there exists a local ordered formula representing c.

We claim that this extension is easily achieved. It will readily be seen by the reader that the basic definitions - s-ball, s-cell complex, s-block bundle, s-cell, etc. - may be mimicked in the context appropriate to the study of A-homology manifolds. For example, an s-ball must now be taken to mean an ordered simplicial

complex constituting an A-homology-manifold with boundary having the

A-homology type of D^j, S^{j-1} for some j. The analogue $qA(n)$ of Q_n, (an s-cell complex in the new sense) is then constructed as is, mutatis mutandis, the appropriate canonical A-homology s-block bundle $\gamma(n)$ over $qA(n)$.

One then readily sees, just as in the proof of 1.6, that given an A-homology manifold M with locally-ordered triangulation T, we obtain a "cellular" map $\tau : T^* \to qA(n)$ classifying the tangent A-homology block bundle of M (here T^* is the dual A-homology cell structure on M^n). τ is, of course, canonical, and, given any co-chain representative $\Gamma \in C^i(qA(n);G)$ of $c(\gamma_A(n)) \in H^i(qA(n);G)$, $\tau^{\#}\Gamma$ represents cM and, since τ is determined locally, we see that $\tau^{\#}\Gamma((\sigma^{n-i})^*,0)$ depends only on the ordered, oriented class of the co-dimension-i n-star $st\,\sigma^{n-i} = \sigma^{n-i} * \ell k(\sigma^{n-i})$ with its orientation 0. I.e. $\tau^{\#}\Gamma$ is determined by a local ordered formula, as required.

As a special application of the remarks above we take note of the following.

<u>1.15 Corollary</u>. In the category of triangulated rational homology n-manifolds, there exist local (unordered) formulae for rational Pontrjagin classes and rational L-classes.

Proof of 1.15 comes by using the trick of Corollary 1.7 on the generalization 1.14 of Theorem 1.6.

We now turn our attention to the question of obtaining <u>explicit</u> local formulae for characteristic classes of triangulated PL manifolds. Of course, we began our discussion with Whitney's well known characterization of w_i by local formula. Thus, as the reader is doubtless aware, attention has, of late, been concentrated on the rational Pontrjagin classes p_i (and to a certain extent on the L-classes).

We briefly refer to the combinatorial formula for p_1 achieved by Gabrielov, Gelfand and Lussik [GGL]. Although the terseness of this paper leaves the matter somewhat obscure, Gelfand's subsequent note [Ge] makes it clear that a local formula for the first rational Pontrjagin class can indeed by specified by enlarging slightly the results of [GGL]. However there are some points worth noting as to the limitations of this particular approach.

(1) As Gelfand's note makes clear, even though the construction of [GGL] leads to a canonical local formula for the rational p_1, this comes only at the expense of an averaging procedure over all choices of "hypersimplicial" data and is thus not in any clear sense a combinatorial <u>algorithm</u> for a local formula. Thus the geometric content of this formula is still quite obscure.

(2) It is well known, of course, that p_1 is a well-defined integral class for PL manifolds (since PL manifolds are smoothable over the 7-skeleton). Theorem 1.6 therefore guarantees the existence of a local <u>ordered</u> formula for p_1 with <u>integral</u> coefficients. There is no hint of such a formula in [GGL].

In any case, the work of [GGL] has been well clarified and publicized by the expository papers of MacPherson [Mac] and Stone [St_1, St_2], so we may omit further details here.

Less well known is Cheeger's work on local formulas for the real L-classes. Details appear in [Ch_2] so that we merely sketch the approach. The starting point is the well-known work of Atiyah, Patodi, and Singer [A-P-S] on the so-called η-invariant of a Riemannian manifold.

Perhaps the easiest way of characterizing this invariant is by use of the canonical Chern-Weil form ℓ_i for the real L-class. We remind the reader that, given a smooth Riemannian manifold M^n, there is a canonical locally-defined 4i-form $\ell_i(M) \in \Omega^{4i}(M)$ with $d\ell_i = 0$ and $[\ell_i(M)] = L_i(M) \in H^{4i}(M;R)$. The η-invariant is defined for oriented Riemannian $(4i-1)$-manifolds M^{4i-1}. We assume M bounds orientably, i.e. $M = \partial W$. (If not, $q \cdot M$ bounds for some q, in which case we may compute $\eta(M)$ as $\frac{1}{q}\eta(q \cdot M)$.) Under this assumption, let W be equipped with a Riemannian metric isometric with $M \times I$ on a collar neighborhood of M. There are then two obvious "candidates" for the signature of the oriented manifold W. The first is merely the usual algebraic-topological signature $\sigma(w)$ of a manifold-with-boundary i.e. the signature of the intersection form in $2i$-dimensional rational homology. The second is given by

$$\int_W \ell_i(W)$$

(The two coincide of course on manifolds without boundary.)

We define $\eta(M)$ as

$$\eta(M) = \int_W \ell_i(W) - \sigma(W)$$

It is an easy exercise to show that $\eta(M)$ is an invariant of M (with its metric and orientation) and reverses sign when the orientation of M is reversed.

The relevance of the η-invariant to the search for local formulas may be suggested by the following "plausibility" argument (based on a distinctly contrafactual hypothesis.) We may consider a triangulated manifold as a metric complex and thus somehow analagous to a smooth manifold endowed with a Riemannian metric. Now suppose that such manifolds acquired as well a "differential $4i$-form," determined locally, analogous to the Chern-Weil form ℓ_i. This would make possible the definition of the η-invariant of a triangulated $4i-1$ sphere, and it is a reasonably straightforward inference, in this hypothetical game, that for an oriented triangulated $4i$-manifold W, the signature of W could then be computed as $\sigma(W) = \sum_v \eta(\ell k\ v, O)$ the sum being taken over all vertices v of the triangulation (and the triangulated spheres $\ell k\ v$ having appropriate orientations O consistent with that of W). This suggests, in turn, that, for an arbitrary triangulated manifold M^n the local formula sssigning to each oriented dual cell $(\sigma^{n-i})^*$, 0 the real number $\eta(\ell k\ \sigma, 0)$ would represent $L_i(M)$.

Returning to reality, however, the construction of "Chern-Weil forms" for triangulated manifolds hardly seems achievable. Nonetheless, the program may still seem plausible since the η-invariant of a smooth Riemannian $4i-1$ manifold has been shown to be intrinsically definable. The celebrated result of Atiyah, Patodi and Singer is that $\eta(M)$ can be directly analytically defined from the spectrum of the Hodge operator of M, without reference to any presumed $4i$-manifold bounded by M.

Cheeger has subsequently shown that the Hodge operator may also be defined on a certain class of varieties which are smooth manifolds away from singular points of a prescribed kind, and which are provided with a consistent family of Riemannian metrics on well-defined smooth strata. Without getting into other examples, quite important in their own right, we single out for special attention, as an example to which Cheeger's constructions apply, triangulated PL manifolds. Here, the strata in question are the simplices themselves, each metrized by the Riemannian structure of the standard simplex, or indeed, by any consistent set of convex-linear metrics. [We note in passing that triangulated rational homology manifolds can also be treated in much the same way.]

Cheeger goes on to show that, for oriented $4i-1$ varieties of this sort, the Atiyah-Patodi-Singer construction goes through, resulting in a well-defined η-invariant. Moreover, there thus arises a local formula for the real characteristic class L_i, viz,

$$\phi(\sigma^{n-4i} * \Sigma^{4i-1}, 0) = \eta(\Sigma^{4i-1}, 0)$$

where the triangulated sphere Σ^{4i-1} is metrized with the standard metric on each simplex. Note that this is a local formula, rather than a local-ordered formula. However, despite its attractive canonicity, this formula has not been made "combinatorial" in the sense that a direct algorithm for computing it purely from the combinatorial data of Σ^{4i-1} is not yet known. Indeed, the traditional difficulties in actually computing the η-invariant of a smooth Riemannian manifold certainly seem to persist in this context as well as far as is currently known.

An additional attempt to construct local combinatorial formulae for rational classes is to be found in the paper of Gabrielov [Gab] and representa a continuation of the methods of [GGL] whose aim is to extend the compass of those methods so as to handle the higher Pontrjagin classes. The earlier paper, in fact, briefly refers to these additional results. We shall briefly outline Gabrielov's work, or rather, an essentially equivalent formulation. Perhaps an initial apology ought to be made for the present treatment. As presented in [Gab], Gabrielov's results seem to be inordinately hard to read or understand. First of all the paper is exceedingly terse. Proofs are completely absent, even in outline. Moreover, there is no attempt to fill in the geometric and intuitive background of the construction, which thus appears to arise out of thin air. Finally, and rather unfortunately, there is a gap in the construction, filled somewhat inconspicuously by an ad hoc hypothesis, which vitiates any claim that the problem of finding local formulae for p_k has been fully solved by these methods. The terseness of the paper itself, and of references to it, has obscured this point.

Nonetheless, Gabrielov's construction, when put in a clear geometric context, has some interesting aspects. At worst it illuminates the essential difficulty which lies at the heart of such an approach. Thus, the present author feels justified in the exposition below.

First of all, we place a small restriction on the kinds of

triangulated manifolds to be studied. Recall (see, e.g. [Wh]) that an n-star $K = \sigma^{n-i} * \Sigma^{i-1}$ is said to be a <u>Brouwer star</u> if and only if there is an embedding $K \subset R^n$ which is convex linear on all simplices. If M^n is a combinatorially triangulated manifold, we say that the triangulation is Brouwer if $st(\sigma)$ is a Brouwer star for every simplex σ. It is a fact that every combinatorial triangulation of an n-manifold has a Brouwer subdivision. Thus the restriction that the manifolds we study are equipped with Brouwer triangulations is seen to be a rather inessential one.

Given a Brouwer n-star $K \cong \Delta^{n-i} * \Sigma^{i-1}$ we let $C(K)$ be the configuration space of K, viz, the space of embeddings $K \subset R^n$, $b_\Delta \to 0$, modulo the action of the general linear group $GL(n;R)$. The topology on $C(K)$ arises when one notices that the space of embeddings has a natural topology viz, as an open subset of $R^{n \cdot p}$, $p=$ # vertices of K, and that $GL(n;R)$ acts freely on this space; thus $C(K)$ with the quotient topology is a manifold of dimension $n \cdot (p-n)$.

Now suppose that τ^j is a simplex of Σ. The complex $st(\Delta * \tau, K) = K_\tau$ is then an n-star and a Brouwer n-star as well. As a technical device, we now insist that Brouwer stars be equipped with linear orderings on vertices. Thus $\Delta^{n-i} * \tau$ may be identified with $\Delta^{n+r-i+1}$ and K_τ may be canonically identified with $\Delta^{n+r-i+1} * \ell k(\tau, \Sigma^{i-1})$ which has a natural inherited ordering. Given a specific linear embedding $f: K, b_{\Delta^{n-i}} \to R^n, 0$, we obtain an embedding $f_\tau = f|K_\tau, - f(b_{\Delta'}): K_\tau, b_{\Delta'} \to R^n, 0$ where $\Delta' = \Delta^{n-i} * \tau \cong \Delta^{n+v-i+1}$. This obviously induces a continuous restriction map $r_\tau: C(K) \to C(K_\tau)$.

To put things into a convenient universal context, we take one i-cell e_K (to be thought of as $c\Sigma^{i-1}$) for each codimension-i ordered Brouwer n-star $\Delta^{n-i} * \Sigma = K$ (e_K = point for $i = 0$). Ranging over all ordered Brouwer n-start K, we assemble these cells into a

C-W complex $\Theta(n)$ as follows: Given $K = \Delta^{n-i} * \Sigma^{i-1}$, $\tau \subset \Sigma^{i-1}$ identify e_{K_τ} with the cell $\tau^* \subset \Sigma$ dual to τ. The identification is made in a canonical way, i.e. there is a canonical PL homeomorphism from $e_{K_\tau} = c(\ell k(\tau,\Sigma))$ to τ^*. $\Theta(n)$ is thus $\bigcup_K e_K$ modulo these identification. The universality of $\Theta(n)$ lies in the fact that given a locally ordered Brouwer-triangulated manifold M^n there is a canonical Gauss map $g_m: M^n \to \Theta(n)$ which for each simplex σ M sends the dual cell σ^* to the cell $e_{st(\sigma,M)} \subset \Theta(n)$.

We now construct a certain space $B(n)$ "over" $\Theta(n)$. $B(n)$ will not (unfortunately!) be a fiber bundle. In fact, if it were (with compact fibre) the construction of explicit local formulae could be made to follow. However, one might think of $B(n)$ as a certain coarse approximation of a fiber space. For each ordered Brouwer star K, let B_K denote $e_K \times C(K)$. Moreover, if τ is a simplex of Σ, and $x \in \tau^*$, we identify $(x,f) \in e_K \times C(K) = B_K$ with $(y, r_\tau f) \in e_{K_\tau} \times C(K_\tau) = B_{K_\tau}$ where y is a point of e_{K_τ} corresponding to x under the natural identification of e_{K_τ} with τ^*. Thus $B(n) = \bigcup_K B_K$ modulo these identifications. It is clear that the projection map $B_K \to e_K$ is compatible with identifications thereby giving rise to a global projection $\pi: B(n) \to \Theta(n)$ (As a point of additional information, we alert the reader that a construction quite similar to that of $\Theta(n)$ and $B(n)$, differing only in minor details, will be made in §8.)

The intuitive reason that $B(n)$ has in some respects the conceptual status of a fiber bundle over $\Theta(n)$ is that the "fiber" at any point $x \in \text{int}\, e_K$ is $C(K)$ which is a "rough approximation" to the space PL/O well known to students of smoothing theory for combinatorial manifolds. However, we shall not attempt to make this analogy any more precise; certainly "fibers" are by no means "constant" from one cell to an adjacent one.

We remark that the "universal" construction $\Theta(n)$ is not strictly necessary. We might merely have constructed a space $B(M)$ "over" M for any locally-ordered Brouwer-triangulated manifold M^n. That is, over each cell σ^* of the dual triangulation of M, we take the space $B_\sigma = \sigma^* \times C(\mathrm{st}\ \sigma, M)$ and glue the various B_σ's so as to respect the various incidence relations of the dual cell structure. We do this by noting that if $\tau^* < \sigma^*$, $\sigma < \tau$, thus $\mathrm{st}\ \tau \subset \mathrm{st}\ \sigma$ and the restriction map from $C(\mathrm{st}\ \sigma)$ to $C(\mathrm{st}\ \tau)$ is defined. Thus, if $x \quad \tau^* \subset \sigma^*$ we identify $x,[f] \in \sigma^* \times C(\mathrm{st}\ \sigma)$ with $x, r_\rho[f] \in \tau^* \times C(\mathrm{st}\ \tau)$, where $\rho \subset \ell k\ \sigma$ is defined by $\tau = \sigma * \rho$.

Putting $\theta(n)$, $B(n)$, $B(M)$ etc. aside for the moment, we consider a manifold M^n PL embedded in R^ℓ. Recall [Wh] that an $(\ell-n)$-dimensional plane $P \subset R^\ell$ is said to be transverse to M at $x \in M$ if and only if there is a neighborhood U of x in M such that for any $y,z \in U$ with $y \neq z$, $y-z \in P$.

Now let $K = \Delta * \Sigma$ be a Brouwer n-star and consider a "general position" embedding $K \quad R^\ell$, by which is meant an embedding linear on simplices and with the set of vectors $\{b_\Delta - v\}$, v vertices K, linearly independent. [Thus $\ell >> n$ in general]. Let N_K denote the set of ℓ-n planes P such that P is transverse to K at b_Δ. (Note N_K depends on the chosen embedding).

<u>1.16 Proposition</u> (Cairns [Cl], Whitehead [Whd]). For a suitable j, $N_K \cong C(K) \times R^j$.

Now consider a locally ordered Brouwer-triangulated manifold M^n and a general position embedding $M \subset R^\ell$, meaning that the embedding is a general position embedding on the star of every simplex. Let $N(M) = (x,P) \in M \times G_{\ell-n,n} \mid P$ transverse to M at $x\}$.

<u>1.17 Proposition</u>. $N(M)$ is homotopy equivalent to $B(M)$.

Without giving a detailed proof we note that this follows from 1.16 which directly implies that $N_{\mathrm{st}\ \sigma}$ is homotopy equivalent to the configuration space $C(st\ \sigma)$. Of course $N(M)$ depends on a particular choice of embedding $M \subset R^{\ell}$ for a particular ℓ, whereas $B(M)$ is clearly intrinsic to M. Our analysis of characteristic classes begins, however, with $N(M)$. Note the obvious projection map $g: N(M) \to G_{\ell-n,n}$ via $(x,P) \to P$.

Consider a rational characteristic class α (for convenience, normal rather than tangential) thought of as an element of $H^*(G_{\ell-n,n}; Q)$. This pulls back under g to $g^*\alpha = \hat{\alpha}(M)$ $H^*(N(M),Q)$. Since α is a rational class, it is defined on the PL manifold M and we claim

__1.18 Proposition__. $\hat{\alpha}(M) = \pi^*\alpha(M)$.

Briefly, this follows from the fact that there is a canonical $(\ell-n)$-vector bundle $\hat{\nu}$ over $N(M)$ whose fiber at $(x,P) \in N(M)$ is the $(\ell-n)$ plane P. Moreover, $\hat{\nu} \cong_{PL} \pi^* \nu(M)$ where $\nu(M)$ denotes the PL normal bundle of M. Since $\hat{\alpha}(M) = \alpha(\hat{\nu})$ 1.18 follows.

In view of 1.18, we may consider, for the moment, a slightly easier problem than that of finding a local formula for a given characteristic class, viz, the problem of finding a local formula which takes into account an embedding of the manifold in some Euclidean space. That is, we may consider local formulae ϕ which assign a number to each triple

$$K = \Delta^{n-i} * \Sigma^{i-1}$$
$$O = \text{orientation on } \Sigma$$
$$e: K \subset R^{\ell}, \text{ a general-position embedding}$$

(with embeddings deemed equivalent if they differ by an action of the Euclidean transformations on R^{ℓ}.)

Given a manifold M_n triangulated by T and an embedding $f: M^n \subset R^{\ell}$ in general position (with respect to T), we thereby obtain $\phi(T,f) \in C^i(T^*,Q)$. To find such a ϕ representing the characteristic class α, it would certainly suffice via 1.18 to have a "locally-defined" section $M \to N(M)$. But remember [Cl], [Wh] that to have any section whatever (whether "locally defined" or not) implies that the manifold M is smoothable. Thus it is clearly hopeless that a "local formula for a section" exist, for that would imply that all PL manifolds are smoothable.

However, a less exacting, and therefore more feasible plan might be to ask for a transfer rather than a section, i.e. a chain-level homomorphism $t: C_*(M) \to C_*(N(M))$ so that the diagram in cohomology

$$H^*(M) \xrightarrow{proj^*} H^*(N(M)) \xrightarrow{t^*} H(M)$$

factors the identity. Here we understand the following: coefficients are in Q or R; $C_*(M)$ is to be the chains on the first barycentric subdivision T' if T or on some finer subdivision while $C_*(N(M))$ need merely be singular chains; the transfer is locally determined, i.e. on σ^*, (a subcomplex of T') $t|C_*(\sigma^*) \to C_*(proj^{-1}(\sigma^*))$ with the homomorphism depending only on the embedding $f|st\ \sigma \subset R^\ell$ and not on the remainder of M.

Were such a transfer map to exist, we would have a local formula (at least in the more restricted sense where an embedding in R^ℓ is part of the data). An easy way of seeing this, at least for real coefficients, is as follows: Since there is a Gauss map $g: N(M) \to G_{\ell-n,n}$, and since the i-dimensional real characteristic class α is canonically represented by a differential i-form ω, $g^*\omega = \omega(M) \in \Omega^i(N(M))$ represents $\alpha(M)$ (in de Rham cohomology). We define, for a star $K = \Delta^{n-i} * \sum^i$, orientation o on $\sum^i$, embedding $e: K \subset R^\ell$ the number

$$\phi\ (K,o,e) = \int_{td^*} \omega(e(K)).$$

Here d^* is the dual i-cell to Δ^{n-i} in K, regarded as the algebraic sum of its simplices appropriately ordered consistent with o, in the first barycentric subdivision of K. It is now obvious that ϕ represents α, since for an embedding $f: M^n \to R^\ell$ of the triangulated manifold M, the co-chain $\phi(M,f)$ is by definition $t^\# \omega(M)$ representing $t^*\alpha(M) = \alpha(M)$.

A slightly more cumbersome way of proving the same thing, (specifically for the kth Pontrjagin class p_k), one which offers more insight into Gabrielov's approach is as follows: Let Q be an arbitrarily-chosen n-plane in R^ℓ and let $V_Q \subset G_{\ell-n,n}$ be defined by $V_Q = \{P \in G_{\ell-n,n} | \dim P \cap Q \geq 2k\}$ Alternatively V_Q may be thought as the set of n-planes $R \in G_{n,\ell-n}$ such that orthogonal projection of R to Q (equivalently, Q to R) has nullity $\geq 2k$. It is well known that V_Q is a submanifold of $G_{\ell-n,n}$ with singularities. Its non-singular part is a submanifold of codimension 4k whose normal bundle is naturally oriented. Intersection with V_Q is precisely the defining formula for the kth <u>integral</u> Pontrjagin class p_k in the sense that, given a smooth manifold $N^n \subset R^\ell$ and a triangulation of N^n such that the Gauss map $g: N^n \to G_{\ell-n,n}$ is in general position with respect to V_Q, then p_k is represented by the cocycle which assigns to each oriented 4k-simplex σ the integer $g(\sigma) \cdot V_Q$.

To extend this slightly, if the ℓ-n bundle ξ over an arbitrary complex X is classified by a specific map $u: X \to G_{\ell-n,n}$, u transverse to V_Q on the 4k-simplices of X, then $p_k(\xi^\perp)$ is represented by the co-chain assigning to each oriented 4k-simplex σ the integer $u(\sigma) \cdot V_Q$. So, in the presence of the (rational or real) transfer $t: C_*(M) \to C_*(N(M))$ we shall have, (for generic Q) the co-chain p assigning to $c \in C_k(M)$ the (rational or real) number $t(c) \cdot g^{-1} V_Q$ where g is now the obvious natural map $N(M) \to G_{\ell-n,n}$. Clearly p is a (rational or real) co-cycle representing the (rational or real) k^{th} Pontrijagin class $p_k(M)$, and, on the assumption that t is locally determined, it is clear that p is, in fact, interpretable as a local formula for p_k.

Of course, the arbitrariness of Q is somewhat unpleasant; we eliminate it simply by averaging over all choices of Q, using the standard $O(\ell)$-invariant measure on $G_{n,\ell-n}$. Thus, the definition of the co-chain p on $C_*(M)$ is replaced by a new definition, equally a local formula, viz

$$(2) \quad p(c) = \underset{Q \in G_{n,\ell-n}}{E} (t(c) \cdot g^{-1} V_Q)$$

where E now denotes expected value over all choices for Q.

We thus see how the construction of a transfer t leads to local formulae for characteristic classes of n-dimensional submanifolds of R^ℓ. We shall show now how to sharpen this formulation so that essentially the same transfer construction generates a local formula in our original sense.

First of all, it is quite obvious that what must be eliminated from the formula above is any specific dependence on the embedding of M^n in R^ℓ and on ℓ. First of all, we note that, given a linear order on the vertices of the triangulation of M^n (say there are ℓ vertices in all), there is a standard embedding of M^n in R^ℓ by extending linearly on simplices the assignment $v_i \to b_i$ where v_i is the ith vertex of M and b_i the ith standard basis vector of R^ℓ. Noting that with this embedding, the formula (2) for p is independent of ordering, we see that the remaining obstacle to an explicit local formula (modulo, of course, the construction of t) is that there remains in our formula a dependence on the number ℓ of vertices, which is global data, albeit of a very weak sort, rather than local. We show how to eliminate this dependence. Given a simplex σ, consider the subspace of R^ℓ spanned by the vertices (say, j of them) of star σ. For the sake of simplicity we may think of this as the standard R^j-space embedded in the standard way in the standard R^ℓ. Let $L(\sigma)$ be the set of ℓ-n planes P transverse to M (i.e. to st σ) at b_σ such that $P^\perp \in G_{n,j-n}$ (i.e. the orthogonal complement of R^j in R^ℓ is a summand of P). Note that $L(\sigma) \times st(\sigma) \subset N(st\ \sigma) \subset N(M)$.

1.19 Proposition. For γ a 4k-chain of $L(\sigma) \times st(\sigma)$

$$\sum_{Q \in G_{n,\ell-n}} \gamma \cdot g^{-1} V_Q = \sum_{Q \in G_{n,j-n}} \gamma \cdot g^{-1} V_Q$$

We omit the proof, which is routine. We now replace $N(M)$, in our considerations, by a smaller space. Recall that given a triangulated manifold there is a more-or-less standard handle-body decomposition with one r-handle h_σ for each r-simplex σ. The diagram below illustrates the situation for a 2-manifold, with the solid lines indicating the simplices and broken lines the corresponding handles

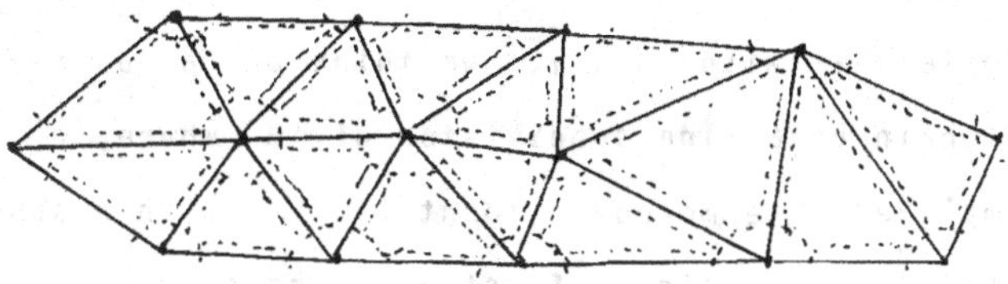

Let $\overline{B}(M)$ be given by $\bigcup_\sigma h_\sigma \times L(st\ \sigma)$. Note that if $\sigma < \tau$, $L(\sigma) \subset L(\tau)$ which tells us what identifications are to be made in forming this union. Note that the construction of $\overline{B}(M)$ as an abstract space depends only on M, not on the standard embedding $M \subset R^\ell$ nor on the ordering of M. $\overline{B}(M)$ is assembled from pieces determined purely by local data. In fact, in analogy to $B(n)$ we may form the space $\overline{B}(n)$ over $\theta(n)$ so that $\overline{B}(M)$ is the "pullback" of the "mock-bundle" $B(n)$ over $\theta(N)$ via the gauss map $M \to \theta(n)$ whenever M is ordered. We further note that there is a canonical homotopy equivalence $\overline{B}(M) \simeq B(M)$; thus $\overline{B}(M) \simeq N(M)$.

Now suppose we had a (rational or real) transfer $t: C_*(\theta_n) \to C_*(\overline{B}(n))$ or, equivalently, locally determined transfers $t: C_*(M) \to \overline{B}(M)$. Then, since we realize $\overline{B}(M)$ as a subspace of $N(M)$ via a standard embedding, we obtain, as in (2) a formula for a real co-chain representing the kth real Pontrjagin class. However, it follows from 1.19 that the value of this co-chain on an orientation of a cell dual to some $n-k$ simplex σ depends on nothing but the combinatorial structure of $st\ \sigma$.

More explicitly, if ρ is an (oriented) simplex of the first barycentric subdivision of the original triangulation, $\rho \subset \sigma^*$, σ an $(n-k)$-simplex, then we may compute

$$(3) \quad p_k(\rho) = \sum_{\sigma \leq \tau} E_Q(t(\rho \cap h_\tau) \cdot V_Q)$$

To explicate, to compute each summand on the righthand side, corresponding to a simplex τ with $\sigma < \tau$, we think of ρ as replaced by the obvious chain on a fine subdivision of M where ρ, $\rho \cap h_\tau$ are all subcomplexes. We moreover identify L_τ with a subspace of $G_{j(\tau)-h,k}$ ($j(\tau)$ = # vertices of $st\ \tau$), and thereby average over $Q \in G_{n,j(\tau)-n}$ with V_Q thought of as a subvariety of $G_{j(t)-n,n}$. Prop. 1.19 is used to show that the result is the same as if the average in each instance were to be taken over all $Q \in G_{n,\ell,n}$. But clearly, $p_k(\rho)$ depends only on ρ and the combinatorial structure of $\ell k\ \sigma$. Hence, taking $p_k(\sigma^*) = \sum_{\rho \subset \sigma^*} p_k(\rho)$ we get a local formula for $p_k(M)$.

Of course, the formula (3) is still not quite satisfactory in that even a rational transfer yields only a local formula with real co-efficients since we are averaging over a non-finite measure space. However, we assert (without proof), the following slightly stronger fact.

<u>1.20 Proposition</u>. In formula (3) above we interpret the term $E(t(\rho \cap h_\tau) \cdot V_Q)$ as an average over the basic n-planes Q (i.e., those spanned by n standard basis vectors) in $R^{j(\tau)}$. The resulting co-chain still represents the Pontrjagin class p_k.

Here, a clear consequence is that a rational transfer leads to a local formula with rational co-efficients.

A final refinement in our analysis is a computation of $t(\rho \cap h_\tau) \cdot V_Q$ where Q is the n-plane spanned, say, by vertices $v_{i(1)} \dots v_{i(n)}$ of the standard $R^{j(\tau)}$. For the sake of simplicity, think of t acting on $(\rho \cap h_\tau)$ as if it were merely a map of the manifold $\rho \cap h_\tau$ to $L_{(\tau)} \times st\ \tau$, (there is no essential new difficulty introduced when treating $t(\rho \cap h_\tau)$ as a singular chain with rational or real co-efficients). Then there is a projection $L(\tau) \times st\ \tau \xrightarrow{\pi} C(st\ \tau)$, the space of configurations of $st\ \tau$, (in fact, of course, this projection has contractible fiber). The assertion $t(x) \in V_Q$ for $x \in \rho \cap h_\tau$ may be characterized in the following way: $\pi t(x)$ is a configuration of $st(\tau)$, i.e. an embedding $st(\tau) \subset R^n$; the vectors $v_{i(1)} \dots v_{i(n)}$ spanning Q are vertices of $st(\tau)$ as well. Thus $\{\pi t(x) v_{i(j)}\}_{j=1,2..n}$ is a set of n vectors in n-space: We claim $t(x) \in V_Q$ precisely when the vectors $\{\pi t(x) v_{i(j)}\}$ have rank $n-2k$.

With this re-computation of $(t\rho \cap h_\tau) \cdot V_Q$ we have essentially obtained Gabrielov's formula [Gab, Prop. 5.1], or at least the (unstated) Corollary of that formula which results from averaging over choices of "hypersimplicial filament."

It would be very gratifying to cap the present analysis by constructing the locally-defined transfer operation t whose existence has been assumed throughout the analysis above. Unfortunately this construction seems really to be the heart of the matter. Gabrielov skirts the difficulty by assuming <u>ad hoc</u> that the triangulations in question all have the property that for all v, $C(st\ \sigma)$ is connected and is rationally $4k$-codim σ connected. This precisely allows the rather trivial construction of a transfer $C_*(M) \to C_*^{\overline{\pi}}(M)$, at least up to dimension $4k$, by wishing away the obstructions to such a transfer map. However, even with these special assumptions, there is no apparent canonicity to the construction, and certainly no clear geometric content. What does seem clear is that something more must be understood about the topology of the spaces $C(\text{st}\ \sigma)$ and the restriction maps $C(\text{st}\ \sigma) \to C(\text{st}\ \tau)$ for $\sigma < \tau$. Such an investigation seems long overdue, considering the early appearance of these spaces and maps in foundational studies in geometric topology. (Recall that Cairn's proof of the smoothability of 4-manifolds reduced to showing merely that $C(K)$ is connected for $K=\Delta^1 * \Sigma^2$. Thus, much of the foregoing is designed as motivation for just such an investigation.

2. Formal links and the PL Grassmannian $\mathcal{G}_{n,k}$

We begin our discussion of $\mathcal{G}_{n,k}$ by introducing the notion of <u>formal link</u> of dimension $(n,k;j)$. If n and k are understood, we merely refer to the dimension of the link as j.

Let U be a $j+k$-dimensional subspace of the standard Euclidean space R^{n+k}. S_U denotes the unit sphere in U (centered at the origin) and D_U the unit disc. If $\Sigma^{j-1} \subset S_U$ is a topological $(j-1)$-sphere, an admissible triangulation is a triangulation of Σ^{j-1} (as a combinatorial manifold) such that:

(a) For each r-simplex σ of Σ^{j-1}, there is a unique $(r+1)$-dimensional subspace of R^{n+k} containing σ.

(b) If $c(\sigma)$ is the convex hull of the vertices of Σ in R^{n+k}, then σ is the radial projection upon S_U of $c(\sigma)$.

(c) The convex structure of σ corresponds to the natural convex structure on $c(\sigma)$ under the aforesaid radial projection.

In particular, this set of assumptions implies that no two points of a simplex are antipodal in S_U.

<u>2.1 Definition</u>. A formal link L of dimension $(n,k;j)$ is a pair (U_L,Σ_L) where U is a $(j+k)$-dimensional subvector space of R^{n+k} and $\Sigma_L \subset S_{U_L}$ is an admissibly-triangulated $(j-1)$-sphere.

Note that formal links of dimension 0 are defined. In this case $\Sigma_L = 0$, and thus a formal link L merely corresponds to a k-plane U_L in R^{n+k}.

Let L be a formal link of dimension $(n,k;j)$, and let v be a vertex of Σ_L. We shall define a new formal link L_v of dimension $(n,k;j-1)$ as follows: Let ρ denote the segment from the origin to v in U. We let U_v be the $(j+k-1)$-plane orthogonal to ρ in U. Let U' be the affine $(j+k-1)$-plane of U parallel to U_v and passing through the midpoint m of ρ. Let S' be a small $(j+k-2)$-

sphere of radius λ in U', centered at m. If σ is a simplex of $\ell k(v,\Sigma_L)$, let $\tau(\sigma) = \sigma * v$ denote the corresponding simplex of one dimension greater in $st(v,\sigma)$. Let $P(\sigma)$ be the union of all rays in U from the origin to points of $\tau(\sigma)$. We claim that if σ_1 is defined as $S' \cap P(\sigma)$ then σ_1 is homeomorphic to σ. If we let σ range over all the simplices of $\ell k(v,\Sigma_L)$, then $\bigcup_\sigma \sigma_1$ forms a simplicial complex Σ_v' in S', with Σ_v' isomorphic to $\ell k(v,\Sigma_L)$. Now map S' to S_U by the homeomorphism $u \to {}^1/_\lambda(u-m)$. Let Σ_v be the image of Σ_v' under this map. Then Σ_v is seen to be an admissibly-triangulated (j-2)-sphere in S_{U_v}. We may thus define L_v as (U_v,Σ_v).

This construction may be extended. If $L = (U_L,\Sigma_L)$ is a j-dimensional formal link, and σ an arbitrary r-simplex of Σ_L, let $v_o \ldots v_r$ be its vertices, ordered in some fashion. Let L_o be the (j-1)-dimensional link L_{v_o}. Clearly there are vertices $v_1^1 \ldots v_r^1$ of $\Sigma_{L_{v_o}}$ corresponding to $v_1 \ldots v_r$. Then set $L_1 = (L_o)_{v_1^1}$, thereby obtaining vertices $v_2^2 \ldots v_r^2$ of Σ_{L_1} corresponding to $v_2^1 \ldots v_r^1$. Continuing in this fashion we obtain $L_{i+1} = (L_i)_{v_{i+1}^{i+1}}$, for $i < r$, and the process terminates with L_r, which is a formal link of dimension $j-r-1$.

<u>2.2 Lemma</u>. L_r is independent of the ordering of the vertices of σ.

We merely sketch the proof. Let b denote the barycenter of σ and let m be the midpoint of the ray from the origin to b in U_L. Let X_σ denote the unique r+1-plane of U_L in which σ lies, and let U_σ be the (j+k-r-1)-plane of U_L orthogonal to X_σ. Let U' be the affine (j+k-r-1)-plane parallel to U_σ and passing through m, and, as before, let S' be a small (j+k-r-2)-sphere in U' centered at m. Given a simplex τ of $\ell k(\sigma,\Sigma_L)$ we set $\rho(\tau) = \tau * \sigma$ $st(\sigma,\Sigma_L)$ and let $P(\tau)$ be the union of all rays to

$\rho(\tau)$. $S' \cap P(\tau) = \tau_1$ is seen to be homeomorphic to τ and $\bigcup_\tau \tau_1$ is a simplicial complex Σ'_σ isomorphic to $\ell k(\sigma,\Sigma_L)$. Once more, the obvious translation followed by dilation identifies S' with S_{U_σ}, and the image Σ_σ of Σ'_σ is an admissibly-triangulated $(j-r-2)$-sphere. Thus we obtain a $(j-r-1)$-dimensional link $L_\sigma = (U_\sigma,\Sigma_\sigma)$. We claim that L_σ is the same as the link L_r defined above, given the ordering of the vertices of Σ. Thus, since L_σ is obviously independent of this ordering, so is L_r. If $K = L_\sigma$ for some simplex σ of Σ_L we say K is incident to L (written $K < L$).

Given a formal link L of dimension j, consider a vertex v of Σ_L. Let v^* denote the dual $(j-1)$-cell to v in Σ_L. (i.e., v^* is the star of v in the first barycentric subdivision of Σ_L.) This is obviously isomorphic as a simplicial complex to the cone on the first barycentric subdivision of $\ell k(v,\Sigma_L)$. On the other hand, the sphere Σ_{L_v} is naturally isomorphic to the complex $\ell k(v,\Sigma_L)$. Thus, we obtain a natural homeomorphism $h(L,v)\colon c\Sigma_{L_v} \xrightarrow{\cong} v^* \subset \Sigma_L$. (In case L is a 1-dimensional link, Σ_L is a 0-sphere and $\Sigma_{L_v} = \emptyset$. In this instance, we interpret the cone construction cX to denote reduced cone on X^+; thus $c\Sigma_v$ is a point and, obviously, $h_{L,v}$ identifies it with $v^* = \{v\}$.)

If σ is a simplex of Σ_L spanned by vertices $v_0 \ldots v_r$ we obtain a chain of inclusions

$$
\text{(1)}\qquad
\begin{array}{ccccc}
cL_r \to \Sigma_{L_{r-1}} & & & & \\
\cap & & & & \\
c\Sigma_{L_{r-1}} & \to & \Sigma_{L_{r-2}} & & \\
 & & \vdots & & \\
 & & c\Sigma_{L_1} & \to & \Sigma_L \\
 & & & & \cap \\
 & & & & c\Sigma_L
\end{array}
$$

where each horizontal map is of the form $h(L_{i-1}, v^{i-1}_{i-1})$. As a corollary of 1.2, we assert that the composite map $c\Sigma_{L_r} \to \Sigma_L$ depends only on the simplex σ, and not on the order of the vertices. We leave this to the reader. Since $L_r = L_\sigma$, in the notation of 1.2 we denote this homeomorphism by $h(L,\sigma)$ and note that it takes cL_σ homeomorphically onto σ^*, the dual cell of σ in Σ. If σ is a simplex of Σ_L, and τ a face of σ, then by a simple extension of 1.2, we see that $L_\sigma = (L_\tau)_{\tau'}$ where τ' is the simplex of Σ_{L_τ} corresponding to the face τ_1 of σ such that $\tau * \tau_1 = \sigma$. If ρ is another face of σ, we get the diagram

$$(2)\qquad \begin{array}{ccccc} & & \Sigma_{L_\tau} \subset c\Sigma_{L_\tau} & & \\ & \overset{h(L_\tau,\tau')}{\nearrow} & & \overset{h(L,\tau)}{\searrow} & \\ c\Sigma_\sigma & & & & \Sigma_L \\ & \underset{h(L_\rho,\rho')}{\searrow} & & \underset{h(L,\rho)}{\nearrow} & \\ & & \Sigma_{L_\rho} \subset c\Sigma_{L_\rho} & & \end{array}$$

and we claim that this diagram strictly commutes.

We may now form a CW complex. First we take one topological j-cell for each j-dimensional link. Think of this cell as $c\Sigma_L$. We take the union of all such, identifying $c\Sigma_{L_\sigma}$ with its image under $h(L,\sigma)$ in $\Sigma_L \subset c\Sigma_L$. We denote this complex by $\mathcal{G}_{n,k}$. The notation is meant to suggest an analogy with the classical Grassmannian $G_{n,k}$, the space of linear n-planes in (n+k)-space. We use the notation e_L to denote the cell of $\mathcal{G}_{n,k}$ which is the image of $c\Sigma_L$.

We now attempt to justify this notational analogy. Consider a triangulated combinatorial manifold embedded, or merely immersed, in R^{n+k}, so that every simplex σ of M^n is linearly embedded (i.e. the image of σ is the convex hull of the images of its vertices).

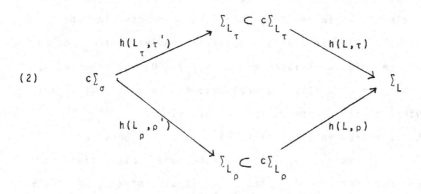

In particular, under such an immersion, the star of every simplex is embedded in R^{n+k}. Let M^n_o denote $\bigcup_\sigma \sigma^*$ where the union is taken over those simplices σ not contained in ∂M^n, and σ^*, as usual, denotes the dual cell of σ. If M^n has no boundary, then $M^n_o = M^n$; if $\partial M^n \neq \emptyset$, then M^n_o is a codimension-0 submanifold of int M^n and differs from M^n merely by a collar neighborhood of ∂M^n.

Given a simplex σ^j of M^n, σ ∂M we assign to it a certain formal link, $L(\sigma,M^n)$ of dimension $(n,k;n-j)$, as follows: Let Y^j_σ be the affine j-plane in R^{n+k} containing σ, and let U_σ be the $(n+k-j)$-plane through the origin orthogonal to Y^j_σ. Let U'_σ be an $(n+k-j)$-dimensional affine plane parallel to U_σ and passing through the barycenter b_σ of σ. Let S' denote a small $(n+k-j-1)$-sphere in U' centered at b_σ. Given a simplex τ of $\ell k(\sigma,M^n)$ let $\rho(\tau)$ denote the simplex $\tau*\sigma \subset st(\sigma,M^n)$, and let $\tau' = S' \cap \rho(\tau)$. Let $\sum' = \bigcup_\tau \tau' \cap S'$. Map S' onto S_{U_σ} by the usual translation-followed-by-dilation $x \to 1/_\lambda (x-b_\sigma)$, λ = radius S', and let $\sum_\sigma$ be the homeomorphic image of $\sum'$. The triangulation of $\sum_\sigma$ whose simplices are $\{\tau'' = \text{image } \tau'\}$ is admissible and is obviously isomorphic to $\ell k(\sigma,M^n)$. We thus obtain a formal $(n-j)$-link $(U_\sigma,\sum_\sigma)$ which is, by definition, $L(\sigma,M^n)$.

This assignment gives rise to a natural map, which we call the Gauss map, $g: M^n_o \to \mathcal{G}_{n,k}$, g depending, of course, on the triangulation of M^n. On the cell level it may simply be described as sending the cell σ^* of M^n_o to the cell $e_{L(\sigma,M^n)}$ of $\mathcal{G}_{n,k}$. The consistency of this assignment with face relations in the respective complexes follows from the observation that if τ,σ are simplices of M^n with τ a face of σ (i.e. σ^* a face of τ^*) then $L_{(\sigma,M^n)} < L_{(\tau,M^n)}$.

For a more specific description of the map g pointwise, think of σ^* as the cone on the first barycentric subdivision of $\ell k(\sigma,M^n)$. There is a natural simplicial isomorphism $\ell k(\sigma,M^n) \to \sum_{L(\sigma,M^n)}$, and thus a canonical homeomorphism $\partial\sigma^* \to \Sigma_{L(\sigma,M^n)}$, which extends to $\sigma^* \xrightarrow{g_\sigma} c\Sigma_{L(\sigma,M^n)}$.

If we compose this with the map $c\Sigma_{L(\sigma,M^n)} \to e_{L(\sigma,M^n)} \subset \mathcal{G}_{n,k}$, we describe $g|\sigma^*$. Again, we assert that the face inclusions on the dual cell complex M^n_o are consistent, under g with the face relations on the cell complex $\mathcal{G}_{n,k}$ not only cell-by-cell, but also point-by-point. The easy verification of this fact is a matter of observing that the diagram

$$\begin{array}{ccc} \sigma^* & \xrightarrow{g_\sigma} & c\Sigma_{L(\sigma,M^n)} \\ \cap & & \downarrow h \\ \tau^* & \xrightarrow{g_\tau} & c\Sigma_{L(\tau,M^n)} \end{array}$$

is strictly commutative, where h is the map $h(L(\tau,M),\rho)$ and ρ is the simplex of $\Sigma_{L(\tau,M)}$ corresponding to ρ_1 in $\ell k(\tau,M^n)$ with $\sigma = \rho_1 * \tau$. This observation is a matter of direct inspection.

The simplicity and naturality of the Gauss map are self-evident. However, to deserve designation as a Gauss map, as the reader will no doubt observe, there should be an equally natural covering by a PL bundle map, just as the Gauss map $g\colon M^n \to G_{n,k}$ of a smooth immersion is naturally covered by a map from the tangent vector bundle TM^n to the canonical n-plane bundle over $G_{n,k}$. We are thus obliged, first of all, to show that there exists a canonical PL n-bundle $\gamma_{n,k}$ over $\mathcal{G}_{n,k}$, and then to show that the Gauss map $g\colon M^n \to \mathcal{G}_{n,k}$ is naturally covered by a PL n-bundle of M^n. Henceforth, we shall use the term Gauss map so as implicitly to subsume this covering bundle map.

We need some further definitions. Given a formal link L, let X_L denote the orthogonal complement of U_L in R^{n+k}. Let $Q_L \subset U_L$

denote the union of all infinite rays in U_L from the origin through points of Σ_L. In particular, $Q_L = \bigcup_\sigma Q_\sigma$, where σ ranges over the simplices of Σ_L and Q_σ is the union of all rays through points of σ. (If L is 0-dimensional, Q_L is understood to mean the origin.) If L is a j-dimensional link, it is clear that Q_L is a piecewise-linear j-plane in U_L R^{n+k}. X_L is, of course, $(n-j)$-dimensional. Thus $Q_L \times X_L$, the vector sum of the sets Q_L and X_L in R^{n+k}, is a piecewise linear n plane in R^{n+k}. We denote this space by V_L.

We now construct a certain "tautological" map $\mathcal{G}_{n,k} \to R^{n+k}$. We begin by first supplementing the natural cell structure on $\mathcal{G}_{n,k}$ by an additional decomposition. We are going to represent $\mathcal{G}_{n,k}$ as the union of contractible subspaces $\bar{e}_L$, one for each formal link L. In general, $\bar{e}_L$ will neither contain nor be contained in e_L. However, if C is any cellular subcomplex of $_{n,k}$, say $C = \bigcup_{i\in\mathcal{I}} e_{L_i}$, for some indexing set $\mathcal{I}$, then $\bar{C} = \bigcup_{i\in\mathcal{I}} \bar{e}_{L_i}$ will contain C as a deformation retract.

To define $\bar{e}_L$, we first define spaces $\bar{e}_{L,*}$ and $\bar{e}_{L,\sigma}$, where σ is a simplex of Σ_L. We may think of these as subspaces of $c\Sigma_L$, since the natural map $c\Sigma_L \to e_L$ will restrict to a homeomorphism on these spaces. Take the second barycentric subdivision of $c\Sigma_L$ (noting that the first subdivision is the simplicial cone on the first subdivision of Σ_L, which we write as $c\Sigma'$, to specify it as a simplicial complex). Call this second subdivision, (i.e. the first subdivision of $c\Sigma_L'$) C_L. Let $\bar{e}_{L,*}$ denote the simplicial regular neighborhood of the cone point in this complex. Let $\bar{e}_{L,\sigma}$ be the regular neighborhood of the barycenter b_σ of the simplex σ in Σ_L. We identify these spaces with their homeomorphic images in e_L.

Now we let $\bar{e}_{L,J} = \bigcup_{\{\sigma | J=L_\sigma\}} \bar{e}_{L,\sigma} \subset e_L$. Finally, we let $\bar{e}_L = \bigcup_{L<K} \bar{e}_{K,L}$.

2.8

We proceed to the definition of the tautological map, which we denote $G: \mathcal{G}_{n,k} \to R^{n+k}$. We shall first define G on each $c\Sigma_L$, viewing the domain as the simplicial complex C_L. We shall then note that this cell-by-cell definition respects the identifications whereby the various $c\Sigma_L$ are amalgamated to form $\mathcal{G}_{n,k}$. Thus, it will be clear that the local definition defines a global map G. Given a vertex v of C_L, note that v must be the cone point $*$ of $c\Sigma_L$, a barycenter b_σ of some simplex σ of Σ_L or else contigous to either $*$ or some b_σ, where "contiguous" means joined by a 1-simplex of C_L. Now if v is $*$ or contigous to $*$, let $\ell(v) = *$. Let $\ell(v) = \sigma$, if $v = b_\sigma$, or if v is contigous to b_σ, but not to $*$, nor to b_τ for any proper face τ of σ. Let $P(*) = \{v | \ell(v) = *\}$, $P(\sigma) = \{v | \ell(v) = \sigma\}$. Note that $P(*)$, together with the various $P(\sigma)$, partition the vertices of C_L into disjoint families. Now let $E(*)$ (resp. $E(\sigma)$) be the subcomplex of C_L spanned by $P(*)$ (resp., $P(\sigma)$), i.e., the largest subcomplex containing no vertices except those in $P(*)$ (resp., $P(\sigma)$). $E(*)$ is naturally isomorphic to the cone on the second barycentric subdivision of Σ_L while $E(\sigma)$, in the same fashion, is isomorphic to the cone on the second barycentric subdivision of $\ell k(\sigma,\Sigma_L) \cong \Sigma_{L_\sigma}$. Thus, for $v \in P(*)$ (resp. $P(\sigma)$), consider $c\Sigma_L$ (resp., $c\Sigma_{L_\sigma}$) as a subspace of R^{n+k}, and let $G(v)$ be the image of v under the composition

$$v \in E(*) \underset{\text{iso.}}{\cong} c\Sigma''_L \underset{\text{homeo}}{\cong} c\Sigma_L \subset R^{n+k}.$$

(resp., "σ", for "$*$" and "L_σ" for "L")

Having defined G on vertices of C_L, we merely extend linearly to all of C_L. But in so doing, we note that this definition of G on each C_L respects the identifications made when $\mathcal{G}_{n,k}$ was formed as the union, mod identifications, of cells $c\Sigma_L = |C_L|$. Therefore, taking the union of all the local definitions of G yields a global map $G: \mathcal{G}_{n,k} \to R^{n+k}$.

It is important to note that $G(\bar{e}_L) \subset V_L$ for all L. This fact allows us to obtain a very simple definition of $\gamma_{n,k}$ "locally." That is, first define $\gamma_{n,k}|\bar{e}_L$ as $(G|e_L)^*TV_L$, where TV_L denotes the tangent PL n-bundle of the PL manifold V_L. We must merely show that these local definitions coincide canonically on intersections of the form $\bar{e}_L \cap \bar{e}_J$. But note that $\bar{e}_L \cap \bar{e}_J \neq \emptyset$ implies that either $L < J$ or $J < L$. So assuming, say, that $L < J$, note that $G(\bar{e}_L \cap \bar{e}_J)$ is contained in the interior of the non-empty n-manifold $V_L \cap V_J$. Thus the bundle $\gamma_{n,k} = \bigcup_L \gamma_{n,k}|\bar{e}_L$ is, in fact, a well defined global bundle over all of $\mathcal{G}_{n,k}$.

Having defined the bundle $\gamma_{n,k}$, we must now show that, for a triangulated n-manifold embedded or immersed in R^{n+k} linearly with respect to the triangulation, the natural Gauss map $g: M_o^n \to \mathcal{G}_{n,k}$ is covered in an equally natural way by a bundle map $TM_o^n \to \gamma_{n,k}$.

We may think of TM_o^n as represented by the assignment to each point $p \in M_o^n$ of a neighborhood U_p of p in M^n so that $\bigcup_p \{p\} \times U_p \subset M_o^n \times M^n$ is a regular neighborhood of $\Delta M_o^n \subset M_o^n \times M_o^n \subset M_o^n \times M^n$. Now consider the given cellular structure on M_o^n, i.e., the cells of M_o^n are the dual cells to the simplices of M^n not contained in ∂M^n. We shall let a_σ denote the dual cell to such a simplex σ of M^n, and then proceed to decompose M_o^n into subspaces $\bar{a}_\sigma$ where $\bar{a}_\sigma$ bears the same relation to σ as $\bar{e}_L$ did to e_L in our prior construction of an alternate decomposition of $\mathcal{G}_{n,k}$. That is, if a_σ is thought of as a certain subcomplex of the first barycentric subdivision M' of M^n, isomorphic to the cone on $(\ell k(\sigma, M^n))'$, then we may subdivide once more, obtaining the triangulation c_σ of a_σ. For σ a face of τ, we define $a_{\sigma,\tau}$ as the regular neighborhood in c_σ of the vertex b_τ, which is the barycenter of the simplex τ of the original triangulation. We let $a_{\sigma,\sigma}$ be the regular neighborhood of b_σ in c_σ. Then $\bar{a}_\sigma = \bigcup_{\tau \leq \sigma} a_{\sigma,\tau}$.

Under this definition, the Gauss map g clearly takes $\bar{a}_\sigma$ to $\bar{e}_{L(\sigma,M^n)}$. In order to construct the PL bundle map $TM^n_o \to \gamma_{n,k}$ covering g, it will suffice to construct maps $TM^n_o|\bar{a}_\sigma \to TV_{L(\sigma,M^n)}$ covering $G \circ g$. Consider, therefore, a point $p \in \bar{a}_\sigma$, and the fiber U_p of TM^n_o at p. Since M^n is thought of as immersed in the standard R^{n+k}, the simplex σ determines an affine plane X_σ of the same dimension, i.e., the unique such plane containing σ. Consider those points q in U_p such that the vector $p-q$ is orthogonal to X_σ. This set constitutes a PL disc W_p of dimension $r = n-\dim \sigma$. If U_p is taken small enough, we claim that $G \circ g$ takes W_p homeomorphically onto a PL r-disc Z_p, with $G \circ g(p) \in Z_p$, so that Z_p is PL-transverse to the affine plane of dimension $= \dim \sigma$ parallel to X_σ and passing through $G \circ g(p)$. Moreover, Z_p lies entirely in $V_{L(\sigma,M^n)}$. We extend the map $W_p \to Z_p$ to a homeomorphism $U_p \to T_p$ where T_p is a neighborhood of $G \circ g(p)$. In particular, if U_p is viewed, (without loss of generality) as $W_p + d$, for d a small r-disc parallel to X_σ, then T_p may be viewed as $Z_p + d$. This set of homeomorphisms may then be taken to define a bundle map $TM^n_o|\bar{a}_\sigma \to \gamma_{n,k}|\bar{e}_{L(\sigma,M^n)}$, covering the Gauss map g.

It remains to observe that this family of bundle maps is consistent, i.e., its elements piece together to form a well-defined PL bundle map $TM^n_o \to \gamma_{n,k}$ covering g. Details are left to the reader as an exercise.

We remark that there is an alternative way of visualizing the bundle $\gamma_{n,k}$ over $\mathcal{G}_{n,k}$ and the bundle map $TM^n_o \to \gamma_{n,k}$ which covers the Gauss map. Consider the spaces $c\Sigma_L$ which are the pre-images of the cells e_L of $\mathcal{G}_{n,k}$. We may replace $c\Sigma_L$, for conceptual purposes, by $b_L \subset R^{n+k}$ where b_L is the linear complex in

R^{n+k} whose vertices are the origin plus the vertices of $\sum_L$ and whose simplices are the convex hulls of those sets of vertices of $c\sum_L$ which span simplices of $c\sum_L$. That is, b_L is the cone on the "inscribed" polyhedron which corresponds to $\sum_L$. Since b_L is isomorphic to $c\sum_L$ as a complex, we may think of b_L, rather than $c\sum_L$ as the preimage of e_L. As a subspace of R^{n+k}, it is clear, $b_L \subset c\sum_L \subset V_L$. Let $\beta_L = TV_L|b_L$. Next, consider a simplex σ of b_L (which corresponds to a simplex of $\sum_L$). Abusing our established notation minimally, we obtain a face L_σ of L and a map $h(L,\sigma): b_{L_\sigma} \to b_L$ which identifies b_{L_σ} with the dual cell to σ in b_L. In fact $h(L,\sigma)$ extends to a map $\psi: V_{L_\sigma} \to R^{n+k}$. That is, think of V_{L_σ} as $Q_{L_\sigma} \oplus X_{L_\sigma}$. Then $h(L,\sigma)$ extends to $\phi: Q_{L_\sigma} \to R^{n+k}$ by $\phi: tp \to h(L,\sigma)(0)+t(h(L,\sigma)(p)-h(L,\sigma)(0))$. Here $p \in \dot{b}_{L_\sigma}$, $t \in R^+$, thus tp is a typical element of Q_{L_σ}. Now extend ϕ linearly to ψ on $Q_{L_\sigma}+X_{L_\sigma}$ by $\psi(q+x) = \phi(q)+x$, $q \in Q_{L_\sigma}$, $x \in X_{L_\sigma}$. Clearly, ψ is a PL embedding. Moreover, in a neighborhood of $h(L,\sigma)(b_{L_\sigma})$ one sees that V_L coincides with $\psi(V_{L_\sigma})$. Thus, we have a standard way of identifying $TV_{L_\sigma}|b_{L_\sigma}$ with $TV_L|h(L,\sigma)b_{L_\sigma}$. In other words, we obtain bundle maps $\theta(L,\sigma): \beta_{L_\sigma} \to \beta_L$ covering $h(L,\sigma)$. Moreover the same consistency conditions hold for the θ's as for the h's, i.e. diagram (2) above will still commute if the $c\sum_L$'s be replaced by b_L's and the h's by θ's. Thus (now viewing $\mathcal{G}_{n,k}$ as having been pieced together from b_L's rather than $c\sum_L$'s), we have a way of piecing together simultaneously a bundle $\gamma_{n,k}$ from the β_L's.

Given an immersed manifold M^n in R^{n+k} and a cell σ^* of M^n_o, it is clear that, over σ^*, the Gauss map arises from a map $i: \sigma^* \to b_{L(\sigma,M^n)}$; i clearly extends to a homeomorphism u from a neighborhood of σ^* in M^n to a neighborhood of $b_{L(\sigma,M)}$ in $V_{L(\sigma,M)}$, and we thus obtain, locally, a standard map $i: TM^n|\sigma^* \to \beta_{L(\sigma,M^n)}$. These maps, in turn, piece together to form the global map $TM^n|M_o^n \to \gamma_{n,k}$.

We briefly note that, in order to show that the two proposed definitions of $\gamma_{n,k}$ essentially coincide we use techniques which imitate the construction of the covering map $TM^n|M_o^n \to \gamma_{n,k}$ of the Gauss map $g: M_o^n \to \mathcal{G}_{n,k}$, as that construction was done using the initial definition of $\gamma_{n,k}$. More precisely, we let γ_1 denote the first proposed version of $\gamma_{n,k}$ (i.e., locally given by G^*TV_L over $\bar{e}_L$) and γ_2 the second (i.e., locally $TV_L|b_L$). We may show that for any pair of formal links $L < K$, if $U_{K,L}$ is the preimage in b_K of $\bar{e}_L \cap e_K$, there is a bundle map $TV_K|U_{K,L} \to TV_L$ covering $U_{K,L} \subset \bar{e}_L \xrightarrow{G} V_L$. (This construction is, essentially, the construction of $TM^n|\bar{a}_\sigma \to TV_{L(\sigma,M)}$ made earlier.) These locally-defined maps fit together consistently to give a global map $\gamma_2 \to \gamma_1$ covering the identity. Thus $\gamma_2 = \gamma_1 = \gamma_{n,k}$.

Having defined the Gauss map and the natural covering bundle map, we may make some further elementary observations, partly to motivate some of the subsequent chapters.

In the first place, it is natural to look for a double sequence in n and k modeled on the familiar one for standard Grassmannians

$$\begin{array}{ccccccc}
 & & \downarrow & & \downarrow & & \\
\to & & G_{n,k} & \longrightarrow & G_{n+1,k} & & \to \\
 & & \downarrow & & \downarrow & & \\
\to & & G_{n,k+1} & \longrightarrow & G_{n+1,k+1} & & \to \\
 & & \downarrow & & \downarrow & &
\end{array}$$

There is, in fact, a natural double sequence

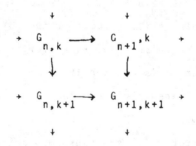

$$(3)\qquad \begin{array}{ccccccc} & & \downarrow & & \downarrow & & \\ \to & \mathcal{G}_{n,k} & \xrightarrow{\alpha} & & \mathcal{G}_{n+1,k} & \to \\ & \beta\downarrow & & & \beta\downarrow & \\ \to & \mathcal{G}_{n,k+1} & \xrightarrow{\alpha} & & \mathcal{G}_{n+1,k+1} & \to \\ & \downarrow & & & \downarrow & \end{array}$$

where α and β are, in fact, inclusions of subcomplexes.

To define α it suffices to define a set map $\mathcal{a}$ from formal links of dimension $(n,k;j)$ to formal links of dimension $(n+1,k,j)$, which is consistent with face relations. Note that a formal link L of dimension $(n,k;j)$ is given by data (U_L,Σ_L) U_L a $(j+k)$-plane in R^{n+k} and Σ_L an admissible triangulated $(j-1)$-sphere in S_{U_L}. But under the standard inclusion $R^{n+k} \subset R^{n+k+1}$, U_L may be considered as a $(j+k)$-plane in R^{n+k+1}, and thus the data (U_L,Σ_L) may be viewed as determining a formal link of dimension $(n+1,k;j)$, which we denote $\mathcal{a}(L)$. Clearly $\mathcal{a}$: $L \mapsto \mathcal{a}(L)$ induces an inclusion $\mathcal{G}_{n,k} \to \mathcal{G}_{n+1,k}$; this is the map α of diagram (3). Of course, we must construct a bundle map $\gamma_{n,k} \oplus \varepsilon \to \gamma_{n+1,k}$ to cover α, if the pattern for the standard Grassmannians is to be followed further. That this bundle map naturally exists may be seen as follows: Note that $V_{\mathcal{a}(L)} \subset R^{n+k+1}$ is just $V_L \times R$, (i.e., $V_L \oplus R_{n+k+1}$, where R_{n+k+1} is a copy of R which is the last summand of $R_1 + R_2 \ldots + R_{n+k+1} = R^{n+k+1}$. Thus $TV_{\mathcal{a}(L)} = TV_L \oplus \varepsilon$ in a natural way. Since $\gamma_{n,k}$ is defined, locally on $\bar{e}_L$, as G^*TV_L, and since $\alpha(\bar{e}_L) \subset \bar{e}_{\mathcal{a}(L)}$, this identification induces the desired bundle map covering α.

As for the definition of β, we once more resort to a set map on the set of formal links. Given the $(n,k;j)$ link $L = (U_L,\Sigma_L)$, think of R^{n+k} as included in R^{n+k+1} in a slightly non-standard way as $R_2 + R_3 \ldots + R_{n+k+1}$, a sum of copies of R. Let $\mathcal{B}(L)$ be given by the data $(U_L \oplus R_1,\Sigma_L)$; $\mathcal{B}(L)$ is a formal link of dimension (n,k+1;j). Once more, $\mathcal{B}$ induces an inclusion of CW complexes $\mathcal{G}_{n,k} \xrightarrow{\beta} \mathcal{G}_{n,k+1}$. Here, however, since it is clear that $V_{\mathcal{B}(L)} = V_L$, as the latter is included in R^{n+k+1} via the above inclusion $R^{n+k} \subset R^{n+k+1}$, and that $\beta(\bar{e}_L) \subseteq \bar{e}_{(L)}$, it follows that $\gamma_{n,k+1}|\mathcal{G}_{n,k} = \gamma_{n,k}$, thus β is covered by a bundle map in the obvious way.

Finally, note that $\alpha \circ \beta = \beta \circ \alpha : \mathcal{G}_{n,k} \to \mathcal{G}_{n+1,k+1}$, so that diagram (3) commutes as does

$$\begin{array}{ccccc} & \downarrow & & \downarrow & \\ \to & \gamma_{n,k} & \xrightarrow{\alpha} & \gamma_{n+1,k} & \to \\ & \beta\downarrow & & \beta\downarrow & \\ \to & \gamma_{n,k+1} & \xrightarrow{\alpha} & \gamma_{n+1,k+1} & \to \\ & \downarrow & & \downarrow & \end{array} \tag{4}$$

(Here we use α's and β's to denote the bundle maps covering the corresponding α's and β's in (3).)

A point which the reader may have noticed is that $\mathcal{G}_{n,k}$ and the Gauss map do not respond at all well to deformation. Thus, if we have a piecewise-linear immersion $m_o : M^n \to R^{n+k}$, and deform it through a continuous family m_t of piecewise-linear immersions (all convex-linear on simplices with respect to a fixed triangulation), we obtain a family of Gauss maps $g_t : M_o^n \to \mathcal{G}_{n,k}$ which is <u>not</u>, in general, continuous. That is, perturbing the "solid angle" which represents the formal link of a simplex in the plane normal to the simplex by even the smallest amount instantly shifts the Gauss map on σ^* very far away.

This suggests a possible retopologization of $\mathcal{G}_{n,k}$ so that e_L and e_K become close when L and K are formal links of the same dimension which are close in an intuitive sense. That is, we may think of $\mathcal{G}_{n,k}$ (i.e., its first barycentric subdivision) as the geometric realization of a simplicial set. The natural retopoligization of $\mathcal{G}_{n,k}$ comes from putting a natural non-discrete topology on the set of j-simplices, for all j, such that face operations are continuous. Thus we may obtain a simplicial space whose geometric realization puts a smaller topology on the underlying point set of $\mathcal{G}_{n,k}$ than the original CW complex. We examine this construction exhaustively in §7 below.

We may note, nonetheless, that even though the $\mathcal{G}_{n,k}$ and Gauss map constructions as originally defined do not behave well under regular homotopy of PL immersions, they do behave well under concordance, at least if we may extend the constructions a bit. Given two PL immersions of M^n into R^{n+k}, say f_0 and f_1, we shall call the two concordant if and only if there is a PL immersion $F: M^n \times I \to R^{n+k} \times I$, with $F^{-1}(R^{n+k} \times \{i\}) = M^n \times \{i\}$, $i = 0,1$, so that $F|M^n \times \{i\} = f_i$. Note that we do <u>not</u> assume that the triangulations on either end coincide. We should like to be able to conclude

<u>2.3 Lemma</u>. If f_0 and f_1 are concordant immersions of M^n into R^{n+k} with respective Gauss maps $g_0, g_1: M_0^n \to \mathcal{G}_{n,k}$ then $\alpha \circ g_0$ is homotopic to $\alpha \circ g_1$ in $\mathcal{G}_{n+1,k}$.

(N.B. The different triangulations of M^n with respect to which the two immersions are linear may give rise to slightly different M_0^n's: nonetheless, these may be identified to all intents and purposes.)

We shall not give a full proof of 2.3. We merely note that it depends essentially on extending the notion of Gauss map to slightly more complicated situations. For the moment, we shall assume that M^n is without boundary; we leave to the reader the minor task of extending the construction below to cover the case where a non-void boundary is present.

Suppose $M^n = \partial W^n$, and suppose further that there is an immertion $f: W^{n+1}, M^n \to R_+^{n+k+1}, R^{n+k}$. Here

$R_+^{n+k+1} = R^{n+k} \times R_+ \quad R^{n+k} \times R = R^{n+k+1}$. We assume that $f^{-1}(R^{n+k}) = M^n$. We wish to define a Gauss map $g\colon W^n, M^n \to \mathcal{G}_{n+1,k}, \mathcal{G}_{n,k}$, where $\mathcal{G}_{n,k}$ is identified with a subcomplex of $\mathcal{G}_{n+1,k}$ via α. This map is to have the property that $g|M^n \to \mathcal{G}_{n,k}$ is the Gauss map of $f|M^n \to R^{n+k}$. Clearly, we may begin by defining g on $W_0^n = \bigcup \sigma^*$ over those simplices σ of W^{n+1} not in M^n. This coincides with the standing definition coming from regarding f as an immersion of W^{n+1} into R^{n+k+1}. We also define $g|M^n$ as the Gauss map of the immersion $f|M^n \to R^{n+k}$. Thus the problem reduces to extending the definition to the collar on $\partial W^{n+1} = M^n$ between M^n and ∂W_0^{n+1}. In effect, this means specifying g on the half-cell $\sigma_W^* =$ dual-half-cell to σ in W for each simplex σ of M^n so as to agree with g as already defined on $\sigma_W^* \cap W_0$ and $\sigma^* =$ dual cell to σ in M^n. For each such σ, of dimension $n-j$, we define a certain formal link of dimension $(n+1,k;j+1)$ denoted by $\hat{L}(\sigma,W)$. Let u denote the vector $(0,0,0,\ldots 0,-1)$ in R^{n+k+1}. As usual, let b_σ denote the barycenter of σ, and let $c_\sigma = b_\sigma + u$. Clearly, the immersion (i.e., embedding) $f|(st(\sigma,W), st(\sigma,M)) \subset R_+^{n+k+1}, R^{n+k}$ extends to an embedding $\hat{f}_\sigma$ of $D_\sigma = st(\sigma,W) \bigcup_{st(\sigma,M)} c\, st(\sigma,M)$ into R^{n+k+1} by sending the cone point to c_σ and extending linearly over $c\, st(\sigma,M)$. We define $\hat{L}(\sigma,W)$ as $L(\sigma,D_\sigma)$ under this embedding. It is clear from inspection that $L(\sigma,M^n)$ is incident to $\hat{L}(\sigma,W)$, i.e. $L(\sigma,M^n) = (\hat{L}(\sigma,W))_c$ (where c is the vertex of $\Sigma_{\hat{L}(\sigma,W)}$ corresponding to c_σ). (Here, be it also noted, we are identifying the (n,k,j) link $L(\sigma,M)$ with its image under A as an $(n+1,k,j)$-link.) Thus $g|\sigma^*$ factors through the diagram $\sigma^* \to c\Sigma_{L(\sigma,M)} \to \Sigma_{\hat{L}(\sigma,W)}$ $c\Sigma_{\hat{L}(\sigma,W)} \to e_{\hat{L}(\sigma,W)} \subset \mathcal{G}_{n+1,k}$. At the same time, it is equally clear that $g|\sigma_W^* \cap W_0$ factors through $\Sigma_{\hat{L}(\sigma,W)}$. I.e., for each simplex τ of $\ell k(\sigma,W) \cap W_0$, $L(\tau,W)$ is incident to $\hat{L}(\sigma,W)$. Thus, there is a standard map $\sigma^*_w \cap W_o \to \Sigma_{\hat{L}(\sigma,w)}$ through which g factors. Denote these respective maps $\sigma^* \to \Sigma_{\hat{L}(\sigma,w)}$, $\sigma^*_w \cap W_o \to \Sigma_{\hat{L}(\sigma,w)}$ by $\hat{g}_1$ and $\hat{g}_2$ respectively. It is clear that im $\hat{g}_1$ and im $\hat{g}_2$ are disjoint j-discs in the j-sphere $\Sigma_{L,\sigma,w)}$. Thus the homeomorphism $\hat{g}_1 \cup \hat{g}_2$ extends to a homeomorphism $g_\sigma : \sigma^*_w \xrightarrow{\text{onto}} c\Sigma_{L(\sigma,w)}$.

Moreover, it may easily be seen, details being left to the reader, that the maps $\hat{g}_\sigma$ may be chosen to be consistent with incidence relations in M^n and $\mathcal{G}_{n+1,k}$. I.e., if $\tau < \sigma$, then $\hat{L}(\sigma,W) < \hat{L}(\tau,W)$ and $h\hat{g}_\sigma = \hat{g}_\tau$ on σ^*_w, where h is the face inclusion $c\Sigma_{\hat{L}(\sigma,W)} \to \Sigma_{\hat{L}(\tau,W)}$.

As to lemma 2.3, its proof may easily be derived from this construction (and slight modifications thereof). I.e., given a concordance $F: M^n \times I; M^n \times 0, M^n \times 1 \to R^{n+k} \times I; R^{n+k} \times 0, R^{n+k} \times I$ of immersions $f_o, f_1: M^n \to R^{n+k}$, one easily produces a Gauss map $G: M^n \times I; M^n \times 0, M^n \times 1 \to \mathcal{G}_{n+1,k}; \mathcal{G}_{n,k}, \mathcal{G}_{n,k}$ which restricts on either end to the Gauss maps g_o, g_1 for f_o, f_1.

3. Some variations of the $\mathcal{G}_{n,k}$ construction

Before we go on to prove some results clarifying the geometric significance of the space $\mathcal{G}_{n,k}$ and what we have called the Gauss map of an immersion, we shall point out that analogous constructions are possible in a variety of geometric contexts. Results analogous to those we shall establish for $\mathcal{G}_{n,k}$ in §4 are available for these spaces as well, but we shall content ourselves with a mere outline of the constructions.

First, we take note of a "Grassmannian" that arises naturally when one considers normal block bundles of triangulated PL submanifolds of triangulated Euclidean space (or, slightly more generally, normal block-bundles of immersions which are simplicial maps with respect to a triangulation of Euclidean space). Historically, this was the first "Grassmannian" devised by the author to deal with the theory of PL immersions. In fact, it was this space which was used to prove a preliminary version of Theorem 1.6 on the existence of local combinatorial formulae for characteristic classes. (C. Rourke's sharpening of this construction for the purposes of dealing with characteristic class questions led to the present version of the result as exposited in Chapter 1 and [Le-R.].) Nonetheless, the block bundle version of the PL Grassmannian is worth studying in its own right. Some results concerning it have appeared in [Le]. These are precisely analogous to the results for $\mathcal{G}_{n,k}$ which will be developed in the subsequent chapter.

The definition of the Grassmannian $\tilde{\mathcal{G}}_{n,k}$ for PL blockbundles (N.B.: This space was denoted $\mathcal{G}_{n,k}$ in $[\tilde{Le}_1]$) proceeds as follows : The general idea is to produce a map $M^n \to \tilde{\mathcal{G}}_{n,k}$ whenever M^n is embedded (or, more generally, immersed) in R^{n+k} as a subcomplex of a triangulation of R^{n+k} whose simplices are convex linearly embedded with respect to the standard linear structure on R^{n+k}. This map is, of course, to have a natural covering by a block-bundle map from the normal block bundle of the embedding to a canonical block bundle $\tilde{\gamma}_{n,k}$ over $\tilde{\mathcal{G}}_{n,k}$.

A <u>formal</u> <u>b-link</u> L of dimension $(n,k;\ j)$ is a triple (U_L, T_L, Σ_L) where U_L is a $(j+k)$-plane of R^{n+k}, T_L a triangulation of the unit $(j+k-1)$-sphere $S_{U_L} \subset U_L$ and Σ_L is a subcomplex of T_L which is a triangulated $(j-1)$-sphere. T_2 is, of course, to be <u>admissible</u> in the sense of §2. (Thus, it is clear that there is a "forgetful map" $(U_L, T_L, \Sigma_L) \mapsto (U_L, \Sigma_L)$ from formal b-links to formal links.

With n,k assumed to be fixed, we denote the dimension of a formal b-link L by a single parameter j. Given such a b-link L of dimension j and a vertex v of Σ_L, we obtain the derived b-link L_v having dimension $j-1$. The construction is clearly analogous to that of §2 for formal links. That is, we let U_{L_v} be the $(j+k-1)$-plane of U_L orthogonal to the segment ρ from 0 to v. Let U' be the affine plane got by translating U_{L_v} so that it passes through the midpoint of ρ, and let S' be some small $(j+k-2)$ pahere in U' centered at this midpoint. We triangulate S' by letting a typical simplex be of the form $S' \cap P(\sigma)$ where: σ is a simplex of $\ell k(v,T_L)$; $\tau = \sigma * v$ is a simplex of $st(v,T_L)$; $P(\sigma)$ is the geometric cone on $\tau(\sigma)$ with the origin of U_L as cone point. This process provides S' with a triangulation isomorphic, as a simplicial complex, to $\ell k(v,T_L)$. Furthermore, we obtain a subcomplex $E' \subset S'$ by letting Σ' be the union of those simplices $P(\sigma) \cap S'$ of S' such that $\sigma \subset \ell k(v,\Sigma_L)$. Σ' is thus isomorphic to $\ell k(v,\Sigma_L)$ under the obvious isomorphism between S' and $\ell k(v,T_L)$. We may then send S' homeomorphically onto the unit sphere $S_{U_{L_v}}$ of U_{L_v} by the obvious translation-followed-by-dilation procedure, and we obtain thereby an admissible triangulation T_{L_v} of $S_{U_{L_v}}$, which has, as a subcomplex, Σ_{L_v} = image Σ'. L_v is thus defined by the triple $(U_{L_v}, T_{L_v}, \Sigma_{L_v})$.

Extending this procedure, in precise analogy to §2 above, we may in fact obtain a derived $(j-\ell-1)$-dimensional b-link L_σ from the j-dimensional b-link L whenever σ is an ℓ-dimensional simplex of Σ_L. We omit details, stressing that the comparable construction in §2 carries over with no difficulties.

As in §2 we obtain, for any simplex σ in Σ_L a map, $h(L,\sigma)$: $c\,T_{L_\sigma}, c\,\Sigma_{L_\sigma} \to T_L, \Sigma_L$, which is a homeomorphism onto its image. Here, we need only note that the cone on the first derived subdivision of T_{L_σ} is isomorphic in a natural way to the dual cell to σ in the cell-structure on S_{U_L} dual to T_L. This isomorphism, moreover, carries the subcone on the subdivison of Σ_{L_σ} to the cell dual to σ in the dual cell structure of Σ_L. With regard to these face maps $h(L,\sigma)$, the same consistency properties hold as for their analogs in §2.

We thus may form a C-W complex $\tilde{\mathcal{G}}_{n,k}$ having one j-cell for each j-dimensional formal b-link. That is, we have, corresponding to L, the j-cell which may be thought of as $c\,\Sigma_L$; we identify $c\,\Sigma_{L_\sigma}$ with its image under $h(L,\sigma)$. Concurrently, we almost automatically obtain a block-bundle $\tilde{\gamma}_{n,k}$ of fibre-dimension k over $\tilde{\mathcal{G}}_{n,k}$. We assemble the total space of $\tilde{\gamma}_{n,k}$ by taking the union of "blocks" of the form D_{U_L} = unit disc of U_L where L is as usual a formal b-link. This union is taken modulo obvious identifications. That, as we have noted, takes $c\,T_{L_\sigma}$ to T_L and is a homeomorphism onto its image. Of course $c\,T$ is to be identified with D_{U_L} in an obvious way so that the total space of $\tilde{\gamma}_{n,k}$ is to be defined as $\bigcup_L c\,T_L$ with $c\,T_{L_\sigma}$ identified with its image in $T_L \subset c\,T_L$ under $h(L,\sigma)$. By abuse of notation, we label this total space $\tilde{\gamma}_{n,k}$.

noting that $\mathcal{G}_{n,k}$ is naturally included as the "0-section" of a k-dimensional block-bundle. If we let e_L denote the cell of $\mathcal{G}_{n,k}$ which is the image of $c\,\Sigma_L$, d_ℓ will denote the restriction of $\gamma_{n,k}$ to e_L, i.e., d_L is the image of $c\,T_L$.

It is then quite straightforward to construct the Gauss map when M^n is an n-manifold which is a subcomplex of a fixed linear triangulation of R^{n+k}. As in §2, we let $M^n_0 = U\sigma^*$, where σ is a simplex of M^n not contained in ∂M. For any such σ of dimension r we are going to define a b-link L of dimension $n-r$. First of all, σ lies in an affine r-plane Y; we let U_L denote the $n+k-r$ plane through the origin orthogonal to Y. Next, let U' be an affine $n+k-r$ plane parallel to U_L and passing through the barycenter of σ. Let S' be a suitably small $(n+k-r-1)$-sphere in U' centered at the barycenter. We obtain a triangulation of S' by letting a typical simplex be of the form $S' \cap (\sigma * \tau)$, where τ is a typical simplex of $\ell k(\sigma, R^{n+k})$. This triangulation of S' contains a certain sub-complex Σ' consisting of those simplices $S' \cap (\sigma * \tau)$ such that $\tau \subset \ell k(\sigma, M^n) \subset \ell k(\sigma, R^{n+k})$. Σ' is, of course, a PL sphere of dimension $n-r-1$.

By virtue of the usual translation and dilation trick, we may send U' to U_L anmd S' to S_{U_L}, obtaining thereby an admissible triangulation of S_{U_L} -- call it T_L -- and with it, corresponding to Σ', a subcomplex $\Sigma_L \subset T_L$. The data (U_L, T_L, Σ_L) define a formal b-link $L(\sigma)$, as intended. Just as in §2, we may now define a Gauss map $g: M^n_0 \to \tilde{\mathcal{G}}_{n,k}$ which has the property that $g: \sigma^* \to e_{L(\sigma)}$.

(In fact, as in §2, $g|\sigma^*$ factors as $\sigma^* \to c\,\Sigma_{L(\sigma)} \to e_{L(\sigma)}$.) We may also note, however, that $\sigma^* \to c\,\Sigma_{L(\sigma)} \to e_{L(\sigma)}$ extends naturally and without difficulty to a map $\bar{\sigma} \to D_{U_{L(\sigma)}} = c\,T_{L(\sigma)} \to d_{L(\sigma)}$ where $\bar{\sigma}$ now denotes the $n+k-r$ cell dual to σ in the triangulation of R^{n+k}. Thus we obtain a map $\bar{g}: \bigcup_\sigma \bar{\sigma} \to \tilde{\gamma}_{n,k}$. Note that, according to

[R-S], $\bigcup_{\sigma} \bar{\sigma}$ is a regular neighborhood of M^n_0 and represents the normal block bundle of the embedded M^n (restricted to M^n_0). That is, each $\bar{\sigma}$ is the k-block of the normal block bundle over the cell σ^*. It is not at all difficult to see that $\bar{g}$ is, in fact, a block-bundle map covering g.

To summarize, then, by using formal b-links, instead of formal links, and $\tilde{\mathcal{G}}_{n,k}$, instead of $\mathcal{G}_{n,k}$, as our candidate for the "Grassmannian" we obtain a space and a block bundle which is the natural classifying space for normal data of submanifolds of a triangulated Euclidean space. It should be added that the Gauss map construction works equally well for an immersion of a triangulated manifold into a triangulated Euclidean space provided that the immersion is simplicial with respect to these triangulations.

Other variants of the basic idea of §2 also bear brief mention. One such arises in connection with triangulated homology manifolds (with respect to appropriate co-efficients). Let A be any abelian group. Consider simplicial complexes which are homology n-manifolds with respect to homology with co-efficients in A. The appropriate analogy to $\mathcal{G}_{n,k}$ in this instance is a complex which we shall denote $A\mathcal{G}_{n,k}$. The space $A\mathcal{G}_{n,k}$ is the natural target of a Gauss map from an A-homology n-manifold embedded or immersed in Euclidean space. In this context "immersed" means that the map into R^{n+k} is linear on simplices and an embedding on the star of any simplex.

To construct $A\mathcal{G}_{n,k}$ we first stipulate that a formal link of dimension i is defined as a pair (U_L, Σ_L) where U_L is, as before, an $(i+k)$-plane of R^{n+k} and Σ_L is, in this instance, an admissibly triangulated subspace of S_{U_L} which is an A-homology manifold with the A-homology type of an $(i-1)$-sphere. As before, given a link L and a simplex σ of Σ_L we may construct the derived link L_σ and the map $h(L,\sigma)\colon c\,L_\sigma \to \Sigma_L$ which identifies

$c\,L_\sigma$ with the A-homology cell dual to σ in Σ_L. Taking the union of A-homology cells $c\,\Sigma_L$ over all links L and making identifications by the maps $h(L,\sigma)$ yields a complex (which is naturally an "A-homology"-cell-complex); this is $A\mathcal{G}_{n,k}$. If M^n is triangulated A-homology manifold PL immersed in R^{n+k} (in the sense given above) we obtain a map $g\colon M^n_0 \to A\mathcal{G}_{n,k}$, where M^n_0 denotes the union of A-homology cells dual to simplices not in ∂M^n.

Despite the naturality of this construction, there remains a certain difficulty in defining a suitable "bundle" over $A\mathcal{G}_{n,k}$ generalizing $\gamma_{n,k}$ over $\mathcal{G}_{n,k}$. This is because the notion of tangent bundle for homology manifolds involves a "block-bundle" type of construction. This difficulty disappears, however, if we seek to study A-homology manifolds embedded (immersed) simplicially in a triangulated Euclidean space. In this case, we may generalize on the model of $\tilde{\mathcal{G}}_{n,k}$, rather than $\mathcal{G}_{n,k}$, to obtain a space $A\tilde{\mathcal{G}}_{n,k}$. Here, the idea is to view formal links as triples (U_L, T_L, Σ_L) where U_L is an $i+k$ plane in R^{n+k}, T_L an admissible triangulation of U_L and Σ_L a subcomplex of T_L which is an A-homology manifold with the A-homology-type of an $(i-1)$-sphere. We then may form $A\tilde{\mathcal{G}}_{n,k}$ which is automatically a subspace of $A\tilde{\gamma}_{n,k}$, the latter being an A-homology k-block-bundle over the former in the sense of [M-M]. For an A-homology manifold M^n embedded (immersed) simplicially in a triangulated R^{n+k}, we get a Gauss map $M^n_0 \to A\tilde{\mathcal{G}}_{n,k}$ and, in this instance, we obtain as well as A-homology block bundle map from the "normal bundle" of M^n to $A\tilde{\gamma}_{n,k}$.

A final generalization derives from the realization that "Gauss maps" exist, in an appropriate sense, for piecewise linear maps of triangulated manifolds into Euclidean space which are neither embeddings nor immersions. The construction is, in a sense, even more straightforward than that for $\mathcal{G}_{n,k}$.

We define a <u>formal</u> <u>star</u> S' of dimension $(n,k;\,j)$ to consist of data $(\Sigma_S, \varepsilon_S, \iota_S)$ where Σ_S is a triangulated PL sphere of dimension $j-1$, ε_S is a linear ordering on the vertices of the complex $\Delta^{n-j} * \Sigma_S$ consistent with the standard ordering on Δ^{n-j}, and ι_S is a function from the vertices of $\Delta^{n-j} * \Sigma_S$ to R^{n+k}. In addition, ι_S is to have the following property: If $\mathcal{I}$ denotes the convex-linear extension of ι_S to a continuous map $\mathcal{I}: \Delta^{n-j} * \Sigma_S \to R^{n+k}$, then $\mathcal{I}(b_\Delta) = 0$, where b_Δ denotes the barycenter of the standard simplex Δ^{n-j}.

Given a formal star $S = (\Sigma_S, \varepsilon_S, \iota_S)$ of dimension $(n,k;j)$ and a vertex v of Σ_S, the derived formal star S_v is easily specified: $\Sigma_{S_v} = \ell k(v,\Sigma_S)$; the ordering ε_{S_v} on $\Delta^{n-j+1} * \Sigma_{S_v}$ is got by noting that the subcomplex $\Delta^{n-j} * st(v,\Sigma_S) \subset \Delta^{n-j} * \Sigma_S$ may be identified with $\Delta^{n-j+1} * \ell k(v,\Sigma_S)$. Here, $\Delta^{n-j} * v$ is identified with the standard Δ^{n-j+1} by using the ordering on $\Delta^{n-j} * v$ induced from ε_S. At the same time, $\Delta^{n-k+1} * \ell k(v,\Sigma_S) = \Delta^{n-j+1} * \Sigma_{S_v}$ also acquires an ordering via restriction of ε_S and we let this be ε_{S_v}. At to the assignment ι_{S_v} taking vertices of $\Delta^{n-j+1} * \Sigma_{S_v}$ to $S^{n+k-1} \cup \{0\}$, this is obtained as follows: As before, let $\mathcal{I}$ denote the convex-linear extension to $\Delta^{n-j} * \Sigma_S$ of ι_S. Let b' be the barycenter of $\Delta^{n-j} * v$ and $p = \mathcal{I}(p')$. Then, for a vertex w of $\Delta^{n-j+1} * st(v,\Sigma_S)$, let $\iota_{S_v}(w) = \iota_S(w) - p'$. Obviously S_v is defined as $(\Sigma_{S_v}, \varepsilon_{S_v}, \iota_{S_v})$.

A straightforward generalization of this construction allows us to define, for any formal star S of dimension $(n,k;\,j)$ and any simplex σ^r of Σ_S, a derived formal star S_σ with $\Sigma_{S_\sigma} = \ell k(\sigma,\Sigma_S)$. Further details of this generalization will be omitted, it being assumed that the reader is by now sufficiently familiar with this theme from previous examples as to be able to fill in the gaps without difficulty.

The notion of formal star, in turn, leads directly to the

construction of the appropriate Grassmannian, which we label $\mathcal{G}^S_{n,k}$, the "S" in the notation being intended to suggest that $\mathcal{G}^S_{n,k}$ is the appropriate target for the Gauss map of an arbitrarily singular PL map $M^n \to R^{n+k}$. Of course, such maps, in their own right have no particular interest as geometric objects which might at first blush make the construction of $\mathcal{G}^S_{n,k}$ seem to be an exercise in futility. Noting only that such an objection is on record, we proceed to the construction, and declare the intention to clarify, subsequently, the non-trivial aspect of the whole proceedings.

The construction of $\mathcal{G}^S_{n,k}$ goes forward in the accustomed vein: If S is a formal star and σ a simplex of Σ_S, then there is a homeomorphism $h(S,\sigma)\colon c\,\Sigma_{S_\sigma} \to \sigma^* \subset \Sigma_S$ where σ^* denotes as usual the cell dual to σ in the cell structure dual to the given triangulation. Thus, we may, as usual, form the disjoint union of spaces of the form $c\,\Sigma_S$, where S is a formal star of dimension $(n,k;\,j)$ n and k being fixed, and then obtain $\mathcal{G}^S_{n,k}$ by identifying $c\,\Sigma_{S\sigma}$ with $\sigma^* \subset \Sigma_S$ under $h(S,\sigma)$.

Quite clearly, any simplex-wise linear map $f\colon M^n \to R^{n+k}$ gives rise to a Gauss map $g(f)\colon M^n \to \mathcal{G}^S_{n,k}$ (or $g(f)\colon M^n_0 \to \mathcal{G}^S_{n,k}$ if M^n has non-void boundary).

Less immediately obvious is the fact that $\mathcal{G}^S_{n,k}$ carries a canonical PL n-bundle $\gamma^S_{n,k}$. We sketch the construction. First of all, given a j-dimensional formal star S, we have an n-bundle over $c\Sigma_S$, viz., $\tau(\Delta^{n-j} * c\,\Sigma_S) \mid c\,\Sigma_S$, where τ denotes ordinary PL tangent bundle. Denote this bundle by γ_S. We claim that if σ is a simplex of Σ_S, then the homeomorphism $h(S,\sigma)\colon c\,\Sigma_{S_\sigma} \to \sigma^* \subset c\,\Sigma_\sigma$ is canonically covered by a PL bundle map $h(S,\sigma)\colon \gamma_{S_\sigma} \to \gamma_S | \sigma^*$. Therefore, as we assemble the complex $\mathcal{G}^S_{n,k}$ by gluing together the cells $c\,\Sigma_S$ via the maps $h(S,\sigma)$, we may simultaneously assemble the total space of $\gamma^S_{n,k}$ by gluing together the various γ_S's by the maps $h(S,\sigma)$. We assert that this construction does, indeed, provide $\gamma^S_{n,k}$ with a PL bundle structure.

It follows, moreover, that for a given simplex-wise convex-linear map $f: M^n \to R^{n+k}$, the resulting Gauss map $g(f): M^n \to \mathcal{G}^S_{n,k}$ is covered by a bundle map $TM^n \to \gamma_{n,k}$.

A final word is appropriate on the motivation for this particular construction. As we have noted, no particular interest attaches to the set of maps $M^n \to R^{n+k}$, per se. But certainly if we envision more restrictive sets of mappings, i.e. mappings satisfying local geometric conditions, it is not hard to see that a corresponding subcomplex $\mathcal{H}$ of $\mathcal{G}^S_{n,k}$ will be defined by this restriction. That is, a map $f: M^n \to R^{n+k}$ which satisfies the presumed restriction will have a Gauss map $g(f): M^n \to \mathcal{G}^S_{n,k}$ whose image lies in $\mathcal{H}$. Conversely, if $g(f)(M^n) \subset \mathcal{H}$, f will satisfy the restriction. (As an example, we might consider immersions of M^n into R^{n+k}; the corresponding subcomplex $\mathcal{H}$ is, roughly speaking, identifiable with the original $\mathcal{G}_{n,k}$ of §2.)

To further clarify the motivation for introducing $\mathcal{G}^S_{n,k}$, we anticipate the results of the next section, where we will chiefly be interested in $\mathcal{G}_{n,k}$ and its subcomplexes. We shall show (Theorem 4.1) that for appropriate subcomplexes $\mathcal{H} \subset \mathcal{G}_{n,k}$, that if a non-closed manifold M^n admits a map $M^n \to \mathcal{H}$ covered by a bundle map $TM^n \to \gamma_{n,k}|\mathcal{H}$, then there will be an immersion $f: M^n \to R^{n+k}$ whose Gauss map has image in $\mathcal{H}$. The immediate point is that the same sort of result holds, mutatis mutandis, for $\mathcal{G}^S_{n,k}$: For suitable subcomplexes $\mathcal{H} \subset \mathcal{G}^S_{n,k}$ (corresponding to local geometric properties of maps), the existence of a map $M^n \to \mathcal{H}$ covered by a bundle map $TM^n \to \gamma^S_{n,k}|\mathcal{H}$ will guarantee the existence of a map $f: M^n \to R^{n+k}$ such that $g(f)\,(M^n_0) \subset \mathcal{H}$ (i.e., f will satisfy the local geometric condition to which $\mathcal{H}$ corresponds.) It is beyond the scope of the present work to state this result precisely or to prove it. Suffice it to say, the methods of proof of theorem 4.1 can be suitably altered and generalized and the reader may find this to be an interesting exercise. The point, then, is that the $\mathcal{G}^S_{n,k}$ construction is not the gratuitous abstraction it might initially have appeared to be.

4. The immersion theorem for subcomplexes of $\mathcal{G}_{n,k}$

As we have seen in §2 above, given a PL immersion $M^n \to R^{n+k}$ of the triangulated manifold M^n, where M^n is combinatorially triangulated and the immersion linear on simplices, we obtain a Gauss map $g: M_0^n \to \mathcal{G}_{n,k}$ covered by a bundle map $TM_0^n \to \gamma_{n,k}$. The converse of this, in a simpleminded sense is easily seen to be true. If M^n is a PL manifold and $g: M^n \to \mathcal{G}_{n,k}$ is covered by a bundle map $TM^n \to \gamma_{n,k}$, then M^n immerses in R^{n+k}. One sees this by noting that the tautological map $G: \mathcal{G}_{n,k} \to R^{n+k}$ is covered by a fiberwise injective map $\gamma_{n,k} \to TR^{n+k}$ thus TM^n becomes a subbundle of $(G \circ g)^* TR^{n+k}$. Applying the Hirsch-Poenaru immersion theorem [H-P], [P], we obtain the desired immersion. (N.B.: if $k = 0$ we must assume M^n to be nonclosed.)

This section aims at considerably strengthening this result. Suppose, for example, that a map $g: M^n \to \mathcal{G}_{n,k}$ is given which satisfies certain restrictions natural from the point of view of geometry. Can we then produce an immersion $M^n \to R^{n+k}$ so that the resulting Gauss map satisfies the same restriction? To make matters more precise, consider the following

<u>4.1 Definition</u>. A subcomplex $\mathcal{H}$ of the CW complex $\mathcal{G}_{n,k}$ is said to be geometric if and only if $e_L \subset \mathcal{H}$ and $V_K = V_L$ implies $e_K \subset \mathcal{H}$, (where L,K are formal links).

To paraphrase, if one link L represents the same local geometry as another, K (note L and K needn't be of the same dimension for this to happen) then e_L is in a geometric subcomplex if and only if e_K is in the subcomplex. We shall give some important examples of geometric subcomplexes following the proof of the main result of this section. First, however, we state that result.

Assume that M^n has no closed components (i.e., each component is either open or has a non-void boundary).

<u>4.2 Theorem</u>. Let $\mathcal{H}$ be a geometric subcomplex of $\mathcal{G}_{n,k}$. Suppose there is a map $f\colon M^n \to \mathcal{H}$ covered by a bundle map $TM^n \to \gamma_{n,k}|\mathcal{H}$. Then there is an immersion $M^n \to R^{n+k}$ such that the resulting Gauss map $g\colon M^n_0 \to \mathcal{G}_{n,k}$ has its image in $\mathcal{H}$.

Proof: Recall, from §2, the definition of $\bar{e}_L$, for each link L. In particular recall that if $\mathcal{H}$ is a subcomplex of $\mathcal{G}_{n,k}$, then $\bigcup_{e_L \subset \mathcal{H}} \bar{e}_L = \bar{\mathcal{H}}$ is a subspace of $_{n,k}$ containing $\mathcal{H}$ as a deformation retract. Thus, in the case at hand, where $\mathcal{H}$ is the geometric subcomplex of the hypothesis of 4.2, we may as well regard f as a map from M^n to $\bar{\mathcal{H}}$.

An easy inductive argument making repeated use of co-dimension one PL transversality arguments shows that, after a slight deformation of f, we may assume that $f^{-1}(\bar{e}_L) = M_L \subset M$ is a codimension zero submanifold of M^n. Note that the interiors of the various manifolds M^n_L are disjoint from one another. It is, moreover, easily seen that $M_L \cap (\bigcup_{K<L} M_K)$ may be assumed to be a codimension-0 submanifold of ∂M_L.

If we compose the map f with the tautological map $G\colon \mathcal{G}_{n,k} \to R^{n+k}$ we see that $G\circ f\colon M^n \to R^{n+k}$ has the property that $G\circ f(M_L) \subset V_L$. Now, by the definition of $\gamma_{n,k}$ and the fact that there is a bundle map $TM^n_0 \to \gamma_{n,k}|\mathcal{H}$ (i.e., $TM^n_0 \to \gamma_{n,k}|\bar{\mathcal{H}}$) covering f, we obtain, for each L with $e_L \subset \mathcal{H}$ a well-defined bundle map $TM_L \to TV_L$.

Moreover, note that $M_L \cap M_K \neq \emptyset$ only if $L < K$, or vice versa. In this case $V_L \cap V_K$ is a non-void n-manifold with boundary, and, without loss of generality, we may assume that $G\circ f\colon M_L \cap M_K \to \operatorname{int}(V_L \cap V_K)$. Now let M^-_L denote the punctured M_L, i.e., M_L with a small open n-disc removed from each component. Let $M^- = \bigcup_{e_L \subset \mathcal{H}} M^-_L$.

We now proceed inductively. Let us assume that $G\circ f|\bar{M}$ has been deformed, keeping each $\bar{M}_L$ in V_L and each $\bar{M}_L \cap \bar{M}_K$ in $\text{int}(V_L \cap V_K)$, to some map h so that, on $N^{(j-1)} = \bigcup_{\dim L<j-1} \bar{M}_L$, h is an immersion. Note that this means, in particular, that for $\dim L < j-1$, $\bar{M}_L$ has been immersed with codimension 0 into the n-manifold V_L. We may assume further that the deformation to h has been covered by a deformation of the family of bundle maps $\{T\bar{M}_L \to TV_L\}$, and that for $\dim L < j-1$, the bundle map $T\bar{M}_L \to TV_L$ is that induced by the codimension-0 immersion. We wish, now, further to deform h so that the same conditions as stated above will continue to hold with $(j-1)$ replaced by j.

Note in this connection that for j-dimensional L, $\bar{M}_L$ maps to V_L under h, the map being covered by a bundle map $T\bar{M}_L \to TV_L$, and moreover, the co-dimension-0 submanifold $\bar{M}_L \cap N^{(j-1)}$ of ∂M is immersed in V_L. In fact, without loss of generality, we may assume that a collar neighborhood of $\bar{M}_L \cap N^{(j-1)}$ in $\bar{M}_L$ is immersed. Now we may apply the Hirsch-Poenaru immersion theorem for PL manifolds [H-P] to deform $h|\bar{M}_L$ to an immersion of $\bar{M}_L$ into V_L. In fact, the deformation may be made rel the collar on $\bar{M}_L \cap N^{(j-1)}$. We note, in this regard, that the validity of this argument depends on the fact that $\bar{M}_L$ may be obtained from the collar on $\bar{M}_L \cap N^{(j-1)}$ by adding successive handles, all of dimension smaller than n.

Furthermore, we may even insure that this deformation of $h|\bar{M}_L$ will keep $\bar{M}_L \cap \bar{M}_K$ in $V_L \cap V_K$ for all K with $L < K$. We see this as follows. For each n-dimensional link K with $L < K$, take a closed regular neighborhood of $\bar{M}_L \cap \bar{M}_K$ (which maps to $\text{int}(V_L \cap V_K)$ under h). Deform the respective restrictions of h to maps which are immersions on each component of such regular neighborhoods, or at least, on the punctured version of such a component. We extend the deformation to all of $\bar{M}_L$, rel a neighborhood of

$\bar{M}_L \cap N^{(j-1)}$. Since $L < J < K$ implies $\text{int}(V_L \cap V_K) \subset \text{int}(V_L \cap V_J)$, this deformation may be chosen to keep $\bar{M}_L \cap \bar{M}_J$ in $\text{int}(V_L \cap V_J)$.

Now consider $(n-1)$-dimensional links K with $L < K$. In what remains of $\bar{M}_L$ after removal of the interiors of the previous regular neighborhoods, pick a regular neighborhood of $\bar{M}_L \cap \bar{M}_K$ and deform once more, rel the subspaces already immersed, to an immersion at least modulo puncturing into $V_L \cap V_K$. Extend the deformation to all of $\bar{M}_L$.

Proceeding in this manner a dimension at a time, we reach a stage where we have immersed, at least modulo multiple puncturing, a regular neighborhood of $\bar{M}_L \cap (\bigcup_{L<K \text{ or } K<L} \bar{M}_K)$. We then immerse, rel what has been previously immersed, the (punctured) remainder of $\bar{M}_L$. Strictly speaking, we have now immersed a multiply-punctured $\bar{M}_L$ through a deformation of h. But this will of course, contain a smaller copy of M_L within itself, and we may now substitute this smaller copy for the original $\bar{M}_L$. Note that the deformation at all stages is made rel $N^{(j-1)}$, and that it may be extended to a deformation of h on all of $\bigcup_{e_J \subset \mathcal{H}} \bar{M}_J$. This deformation preserves the property that $\bar{M}_J$ goes to V_J and $\bar{M}_J \cap \bar{M}_K$ to $\text{int}(V_J \cap V_K)$ for all links J, K. Finally, the deformation on each $\bar{M}_J$, is covered by a deformation of the bundle maps $T\bar{M}_J \to TV_J$, and, on those $\bar{M}_L$, $\dim L < j$ which are immersed in the respective V_L, the bundle map is that of the immersion.

Proceeding with the induction on j, we reach a stage where each $\bar{M}_L$ is immersed in V_L and thus $\bigcup_{e_L \subset \mathcal{H}} \bar{M}_L = N^{(n)}$ is immersed in R^{n+k}. But note that $N^{(n)}$ is merely M^n with, at most, a countable number of discs removed. Therefore $N^{(n)}$ contains a homeomorph M_1 of M^n itself. Thus M_1 (i.e., M^n) certainly immerses, via restriction, in R^{n+k}. Now set $M_L' = \bar{M}_L \cap M_1$. Without loss of generality, each M_L' may be assumed to be a codimension-0 submanifold of M_1. Now triangulate M_1, so that the immersion is linear, and so that each M_L' is a subcomplex. Thus, given a simplex σ of the triangulation, it must lie in one such M_L' and thus its image under the immersion in the corresponding V_L. One possibility is that some interior point of σ maps, under the immersion, into $X_L \subset V_L$. In this case, it is immediate that $V_{L(\sigma,M_1)}$ coincides with V_L and hence, since $\mathcal{H}$ is assumed to be a geometric subcomplex, $e_{L(\sigma,M_1)} \subset \mathcal{H}$.

On the other hand, suppose no interior point of σ goes into X_L. Let τ be a simplex of $\sum_L$ and let E_τ denote the subspace of V_L consisting of $n+k$ vectors w of the form $w = x+st$ where $x \in X_L$, $t \in \overset{\circ}{\tau}$ and $s > 0$. Then the collection $\{E_\tau\}$ together with X_L, covers V_L. Therefore, if no interior point of σ goes into X_L, there must be a τ such that E_τ contains the image of an interior point of σ. But in this case it is immediate that $V_{L(\sigma,M_1)} = V_{L_\tau}$, and again, by the fact that $\mathcal{H}$ is geometric, we see that $e_{L(\sigma,M_1)} \subset \mathcal{H}$.

Thus, for any σ of M_1, $e_{L(\sigma,M_1)} \subset \mathcal{H}$, and it follows that the Gauss map for the given triangulation and immersion has its image in $\mathcal{H}$. This completes the proof of 4.2.

We shall give some applications of this result, but first we pause for some observations.

(1) Even assuming that the codimension k of the putative immersion is positive, the restriction that M^n have no closed components seems unavoidable if the proof is to go through. The catch seems to be that the immersion $M^n \to R^{n+k}$ is constructed as the union of codimension-0 immersions of pieces of M^n into the various n-manifolds V_L. To get these small immersions, we must avoid

n-handles. This ultimately accounts for the aforesaid restriction. In fact, our result is certainly "best possible" in this sense. We may see this by the following counterexample: Let k be arbitrary, and let W denote an n-dimensional vector subspace of R^{n+k}. Let $\mathcal{H} = \bigcup_{V_L = W} e_L$. $\mathcal{H}$ is clearly a geometric subcomplex. Now if M^n is a connected manifold immersed so that the Gauss map has image in , it is clear that the image of M^n must lie in some affine n-plane of R^{n+k} parallel to W. Hence, M^n may not be closed. On the other hand, any closed parallelizable manifold M^n will admit a map (e.g., the trivial map) to $\mathcal{H}$ covered by a bundle map $TM^n \to \gamma_{n,k}|\mathcal{H}$. Thus, 4.2 cannot possibly hold if the restriction that the manifold in question have non-closed components be removed.

In this regard, we may consider a slightly more interesting counterexample. Let $\mathcal{G}^{(j)}_{n,k}$ denote the smallest geometric subcomplex of $\mathcal{G}_{n,k}$ containing the j-skeleton of $\mathcal{G}_{n,k}$. $\mathcal{G}^{(j)}_{n,k} = \bigcup e_L$ such that $V_L = V_K$ for some j-dimensional link K. The image of the Gauss map of an immersion of M^n in R^{n+k} will lie in $\mathcal{G}^{(j)}_{n,k}$ precisely when no point of M^n has a neighborhood which is "crinkled" by the immersion worse than neighborhoods of points in the interiors of $(n-j)$-simplices.

For example, let $n = 2$ and $k = 1$. An immersion whose Gauss map goes to $\mathcal{G}^{(0)}_{2,1}$ must have its image in some 2-plane; if the image lies in $\mathcal{G}^{(1)}_{2,1}$, then the immersion admits edges where, locally, two planes meet at an angle, but there are no "sharp" points.

We claim that 4.2 cannot hold for the subcomplex $\mathcal{H} = \mathcal{G}^{(j)}_{n,k}$, $j < k$, in the absence of the requirement that M^n have non-closed components. This comes by way of the following interesting fact: Let us define a link L of dimension $(n,k;n)$ as "sharp" if $V_L \neq V_K$ for all K of dimension $< n$.

<u>4.3 Lemma</u> (D. Stone). Let $M^n \to R^{n+k}$ be an immersion where M^n is closed. Then there are at least (n+2)-distinct vertices v of M^n such that L(v,M) is sharp.

Proof: Let Y be a piecewise-linear subspace of R^n (not necessarily a manifold), and $y \in Y$. We shall use the notation T_yY to refer to the tangent cone of Y at y, i.e. the smallest Euclidean cone containing all vectors (w-y) where $w \in W$ and W is some sufficiently small neighborhood of y in Y. Note that for W sufficiently small, this cone is independent of W and hence T_yY is well-defined.

Now let $f: M^n \to R^{n+k}$ be an immersion linear with respect to some triangulation of M^n, M^n closed. For L a link of dimension (n,k;j) let Y_L denote the largest vector space in the cone V_L. Note that L is n-dimensional and sharp if $\dim Y_L = 0$.

Now let U be an n+1 plane in R^{n+k} and π orthogonal projection of R^{n+k} on U. Generically $\pi|Y_{L(\sigma,M)} \to U$ is nonsingular for all simplices σ of M^n. Let $A \subset U$ be the image πfM. A may be triangulated as a PL subset of U so that πfv is a vertex of A for all vertices v of M^n. Now let C denote the convex hull of A. There are at least (n+2)-extremal points in C, i.e. points $c \in C$ which admit tangent cones T_cC containing no non-trivial vector subspaces. Now these extremal points must, in fact, be points of A, i.e. $c = \pi fm$, $m \in M$, for each such C. However, $T_cC \supset \pi V_{L(\sigma,M)} \supset \pi Y_{L(\sigma,M)}$ where $m \in \operatorname{int} \sigma$. Thus, since C is extremal m is not in the interior of any simplex of positive dimension, and hence m must be a vertex v. Moreover, $\dim Y_{L(v,M)}$ cannot be positive, for then T_cC would contain the nontrivial vector space $\pi Y_{L(v,M)}$. Thus, for each c, v must be sharp. Since there are at least (n+2) such c, there are at least (n+2) sharp v's, as required for 4.3.

Thus, there can be no immersion of a closed manifold in R^{n+k} whose Gauss map has image in $\mathcal{G}^{(j)}_{n,k}$ $j < n$.

(2) The requirement in 4.2 that the subcomplex $\mathcal{H} \subset \mathcal{G}_{n,k}$ be geometric, rather than merely arbitrary, is involved in the proof of 4.2 through the fact that even though we get an immersion $M^n \to R^{n+k}$ which is the union of codimension-0 immersions $M'_L \to V_L$. We see, upon triangulating M^n with the M'_L as subcomplexes, that the formal link $L(\sigma, M^n)$ of a simplex σ in M'_L is not necessarily L, or L_τ for some simplex τ in $\int_L$. Rather, $L(\sigma, M^n)$ only resembles L or L_τ in having $V_{L(\sigma,M^n)} = V_L$ or V_{L_τ}. Thus, if we were to start out with a <u>non-geometric</u> subcomplex $\mathcal{H}$, we would obtain an immersion the image of whose Gauss map lies in the smallest geometric subcomplex containing $\mathcal{H}$.

On the other hand, the condition that $\mathcal{H}$ be a geometric subcomplex is a rather natural one, as may be seen from some examples below as well as the previous examples $\mathcal{G}^{(j)}_{n,k}$. If we want to restrict to a class of PL immersions where restrictions depend only on the local geometry of tangent cones, and not on triangulations, then such a restriction corresponds naturally to a geometric subcomplex of $\mathcal{G}_{n,k}$

(3) The diagram

$$\begin{array}{ccccc}
 & \downarrow & & \downarrow & \\
\to & \mathcal{G}_{n,k} & \xrightarrow{\alpha} & \mathcal{G}_{n+1,k} & \to \\
 & \beta\downarrow & & \beta\downarrow & \\
\to & \mathcal{G}_{n,k+1} & \xrightarrow{\alpha} & \mathcal{G}_{n+1,k+1} & \to \\
 & \downarrow & & \downarrow &
\end{array}$$

naturally suggests the conjecture that $\lim_{n,k} \mathcal{G}_{n,k} = BPL$ as the analagous limit $\lim_{n,k} G_{n,k} = BO$ in the classical case. However, it will not be possible to prove this by our methods, and it is probably not true. Of course, there is a natural map $\lim_{n,k} \mathcal{G}_{n,k} = \mathcal{G} \xrightarrow{b} BPL$ arising from the classifying maps $\gamma_{n,k} \to BPL(k)$ of the natural bundles $\gamma_{n,k}$. It is easily seen, that, for any finite complex K, $b_*[K,\mathcal{G}] \to [K,BPL]$ is surjective, (and so, b induces epimorphisms of homotopy groups). To prove this note that a map K to BPL, i.e. a stable PL bundle over K, is represented by the tangent bundle of some PL manifold M^n homotopy equivalent to K. (See [Wa].) Of course, M^n immerses in R^{n+k} for sufficiently large k, with Gauss map $g: M^n \to \mathcal{G}_{n,k}$. Thus $b \circ g$ is, essentially, the original map $K \to BPL$ up to homotopy.

The difficulty in proving b_* a bijection arises from a certain weakness in the conclusion of 4.2. The alert reader will have noted that in 4.2, the Gauss map $g: M^n \to \mathcal{H}$ of the immersion obtained is not asserted to be homotopic in $\mathcal{H}$ (nor even in $\mathcal{G}_{n,k}$) to the original map $f: M^n \to \mathcal{H}$ of the hypothesis of 4.2. The reason for the failure of this homotopy of maps to emerge is closely related to the reasons, adduced above, for imposing the requirement that $\mathcal{H}$ be geometric. That is, if the immersion produced by the proof, $M^n \to R^{n+k}$, the union of codimension-0 immersions $M'_L \to V_L$, allowed of a triangulation of M^n so that, under the Gauss map g, M'_L is sent to e_L, then it would follow that g is homotopic to f. As we have already remarked, however, we do not quite get this. The arbitrary imposition of a triangulation, perturbs, so to speak, the natural association of M'_L with e_L, and it is difficult to recover this even up to homotopy.

This point will be further considered at the end of §7.

We shall briefly look at some further examples of geometric subcomplexes of $\mathcal{G}_{n,k}$. They come by way of D. Stone's work [St3, St4] on piecewise linear metrics on combinatorial manifolds. Although such a metric may be defined more abstractly, the examples to keep in mind come from PL immersions (or embeddings) of a PL manifold M^n

in Euclidean space. For each point in such a manifold, Stone defines the notion of curvature, or rather two notions which correspond, roughly, to minimum and maximum sectional curvature in the classical case of smooth Riemannian manifolds. Given a triangulation of M^n with respect to which the given immersion is linear on simplices, these parameters, $K_+(p)$, $K_-(p)$ for $p \in M^n$ depend only on the simplex σ in whose interior p lies. In fact, they only depend upon the formal link $L(\sigma, M^n)$.

We may define $K_+(L)$, $K_-(L)$ for a formal link L as follows. First of all $K_+(L), K_-(L) = 0$ for L of dimension 0 or 1. Given an arbitrary L, consider the Euclidean cone Γ_L which is the union of all infinite rays in U_L from the origin through points in Σ_L. Let F_L denote the maximal vector subspace of Γ_L. If $F_L = \Gamma_L$, we set $K_+(L), K_-(L) = 0$. If not, let $\hat{U}_L$ be the orthogonal complement of F_L in U_L, and let $\hat{\Sigma}_L = S_{\hat{U}_L} \cap \Gamma_L$. (We do not insist on any particular triangulation of $\hat{\Sigma}_L$.) For $x \in \hat{\Sigma}_L$, let $\mathcal{C}^0(x)$ be the set of points $y \in \hat{\Sigma}_L$ such that there are at least two paths from x to y in $\hat{\Sigma}_L$ of minimal length; let $\mathcal{C}(x)$ be the closure of $\mathcal{C}^0(x)$. Define $K_+(L,x)$ $K_-(L,x)$ as follows: If $\mathcal{C}(x) = \emptyset$, $K_+(L), K_-(L) = 0$. If $\mathcal{C}(x) \neq \emptyset$ then

$$K_+(L,x) = \max_{y \in \mathcal{C}(x)} (2\pi - \cos^{-1}(x \cdot y))$$

$$K_+(L,x) = \min_{y \in \mathcal{C}(x)} (2\pi - \cos^{-1}(x \cdot y))$$

Finally, let $K_+(L) = \min_{x \in \Sigma_L} K_+(L,x)$ and $K_-(L) = \max_{x \in \Sigma_L} K_-(L,x)$.

It turns out that $K_+(L) > K_-(L)$.

Let $a > 0 > b$. We may define a certain geometric subcomplex S_b^a of $\mathcal{G}_{n,k}$ corresponding to immersions of n-manifolds into R^{n+k} such that at each point $a > K_+ > K_- > b$. That is, we let e_L be in the subcomplex S_b^a if $a > K_+(L)$, $K_-(L) < b$ and

$$a > K_+(L_\sigma),\ K_-(L_\sigma) < b$$

for all simplices σ of Σ_L.

In this case, our theorem gives

<u>4.4 Corollary</u>. The non-closed manifold M^n immerses in R^{n+k} such that at each point $a > K_+(p) > K_-(p) > b$ if and only if there is a map $f: M^n \to S_b^a$ covered by a bundle map $TM^n \to \gamma_{n,k}|S_b^a$.

A further corollary can be obtained from the proof, rather than the statement, of 4.2. It concerns characteristic classes, and, in particular the L-classes.

First of all, we note that since $\mathcal{G}_{n,k}$ carries the PL bundle $\gamma_{n,k}$ the rational characteristic classes for any PL manifold M^n will be induced from the corresponding classes for $\gamma_{n,k}$, by the Gauss map $g: M^n \to \mathcal{G}_{n,k}$ of any immersion of M linear on the simplices of some triangulation. So, in particular, if we pick a real cellular co-chain c on $\mathcal{G}_{n,k}$ representing, say $L_k(\gamma_{n,k}) \in H^{4k}(\mathcal{G}_{n,k};R)$, we have the cellular co-chain (on the cell-structure dual to the triangulation) $g^\# c$ representing $L_k(M)$.

Suppose, now, that we have, in addition to the given triangulation T of M, an additional cell-decomposition of M as a regular CW-complex. We assume that these cells are PL subspaces with respect to the original triangulation and are also in general position, i.e. the cells of the CW-complex and the simplices of T are, pair-wise, transverse to each other.

It follows that if a is a cocycle on T^* representing some cohomology class $[a] \in H^i(M;R)$, then we can get a cocycle $\hat{a}$ representing the same class with respect to the cell structure S by virtue of the following formula: Let s be an oriented i-cell of S. Pick orientations of σ^* for all (n-i)-simplices σ of T.

Thus the intersection number $s \cdot \sigma$ and the value of $a(\sigma^*)$ become defined without ambiguity as to sign. Then the assignment

$$\hat{a}: s \to \sum_{\sigma} (S \cdot \sigma)\, a\, (\sigma^*)$$

defines an i-cochain a on S which is clearly a cocycle representing the same underlying cohomology class in $H^i(M;R)$.

We now note Cheeger's work in defining a local formula for the L-classes. This was accomplished, as was mentioned in §1, by extending the Atiyah-Patodi-Singer η-invariant from smooth Riemannian manifolds to more general stratified spaces. In particular, if we consider a formal link of dimension $i = 4k$, $L = (U_L, \sum_L)$, then the value of η on $\sum_L$ (with an orientation on $\sum_L$) gives rise to an oriented co-chain ℓ_i on $\mathcal{G}_{n,k}$ which is a cocycle and which represents, in fact, the characteristic class $L_i(\gamma_{n,k}) \in H^i(\mathcal{G}_{n,k};R)$. ℓ_i is defined on an orientation of the cell e_L as $\eta\sum_L$, $\sum_L$ having the corresponding orientation. Some properties of ℓ_i include the following

(i) $\ell_i(e_L)$ depends only on V_L, and not on L itself. I.E., the metric geometry of $\sum_L$ completely determines the value of η, the combinatorics of $\sum_L$ being irrelevant.

(ii) Suppose $V_L = V_K$ for $\dim K < \dim L$. Then $\ell_i(e_L) = 0$. This is because the assumption that $V_L = V_K$ for some lower dimensional link K means that $\sum_L$ admits an orientation-reversing symmetry.

(iii) Suppose $i < n$. There is a number $\varepsilon > 0$ so that for any t, $0 < t < \varepsilon$, there is an L with $\ell_i(e_L) = t$ (for an appropriate orientation). The proof will not be given.

Thus, if the triangulated manifold M^n is immersed simplex-wise linearly in R^{n+k} via f, we have a well-defined co-cycle, $\ell_i(M,f)$ representing $L_i(M)$. If, in addition to the triangulation, M is given a transverse regular cell decomposition S, we obtain

$\hat{\ell}_i(M,f) \in Z^i(S;R)$ representing $L_i(M)$.

<u>4.5 Lemma</u>. The cocycle $\hat{\ell}_i(M,f)$ depends only on the immersion f and not on the specific triangulation of M.

Proof: Suppose we are given two triangulations T, T' of M such that the given immersion is simplex-wise linear with respect to both, and that the cells of S are in general position with respect to both. Consider a cell S^{4i} of S, with a preferred orientation. The value $\ell_i(M,f)(s)$ is seen to be determined as follows, using T as the triangulation in question: There are a finite number of points $p_1,\ldots,p_m \in S$ such that $p_j \in \sigma_j^{n-4i}$, σ_j^{n-4i} a simplex of T, and $\ell_i(e_{L_j}) \neq 0$, where L_j denotes the formal link $L(\sigma_j, M^n)$. $\hat{\ell}_i(M,f)(s)$ is thus seen to be $\sum_{j=1}^{m} n(\Sigma_{L_j}) = \sum_{j=1}^{m} \ell_i(e_{L_j})$, where each Σ_{L_j} (or e_{L_j}) is given the orientation induced by that of S. If we examine the alternative triangulation T', we see that, for each p_j, $p_j \in \hat{\sigma}_j^{n-4i}$ where $\hat{\sigma}_j$ is a simplex of T'. This follows because, by general position p_j cannot lie in a simplex of T' of dimension $< n-4i$. Moreover, if it lay in a simplex τ of dimension $> n-4i$, then, setting $L_j' = L(\tau,M)$, we would have $V_{L_j} = V_{L_j'}$ and therefore $n(\Sigma_{L_j}) = 0$. So, in fact p_j lies in some $(n-4i)$-dimensional τ with $V_{L'_j} = V_{L_j}$ and hence $\ell_i(e_{L'_j}) = \ell_i(e_{L_j})$ (for orientations consistent with that on S). It follows that $\sum_j \ell_i e_{L'_j} = \sum_j \ell_i e_{L_j}$ (for there will be no additional points of s in n-$4i$-simplices τ of T' with $\ell_i(e_{L(\tau,M)}) \neq 0$). Hence $\hat{\ell}_i(S)$ is independent of triangulation.

We now come to the main application of the technique used in the immersion theorem.

<u>4.6 Theorem</u>. Let M^n be a non-closed manifold for which $L_i(M) = 0$ and suppose that that M immerses in R^{n+k} via f. Let S be a

regular cell decomposition of M. Then f may be modified to an immersion $f_1: M \to R^{n+k}$ such that the induced co-cycle $\hat{\ell}_i(M,f_1)$ vanishes.

Proof: We look at the Gauss map $g(f): M_0^n \to \mathcal{G}_{n,k}$. We put S in general position with respect to a triangulation T of M for which f is simplex-wise linear. We may assume, for simplicity that if s is a $4i$-cell of S and σ an $(n-4i)$-simplex of T, then s meets σ, if at all, at the barycenter of σ. Given this data, we have the $4i$-cocycle $\hat{\ell}_i(M,f)$ on S which, by assumption, is a coboundary. In particular choose c so that $\hat{\ell}_i(M,f) = \delta c$, $c \in C^{4i-1}(S;R)$. Let r be a $(4i-1)$-cell of s, $r \subset M-\partial M$, so that $c(r) = v_r > 0$ (for some orientation of r). Pick a point x_r in r and let D_r be a small n-disc about x. Pick a deformation-retraction of D_r to x with the property that the deformation keeps $\bar{a} \cap D_r$ inside $\bar{a} \cap D_r$ for all closed cells $\bar{a}$ to which r is incident. Picking $x_r \in D_r$ together with this deformation retraction for all r, we have a deformation retraction $\bigcup_r D_r$ to $\bigcup_r \{x_r\}$ (we may assume the D_r are mutually disjoint). We may extend this to a deformation of the identity on M, and, in fact, require further that this deformation preserves closed cells of S and, furthermore, is the constant deformation on: (a) the $4i-2$ skeleton of M, and (b) a neighborhood of $\bigcup_y (g(f))^{-1}y$, where y ranges over all points y in $\mathcal{G}_{n,k}$ where y is of the form: cone pt of $c\Sigma_L \in e_L$, L a formal link of dimension $4i$. If we let ω denote the last stage of this deformation, we obtain a map $g' = g(f) \circ \omega$: $M_0^n \to \mathcal{G}_{n,k}$, and obviously $g' \sim g(f)$. We now further deform g'. The deformation will be rel $M - \bigcup_r D_r$. For each r, pick a formal $4i$-dimensional link K_r and a positive integer k_r so that $k_r \cdot \ell_i(e_{K_r}) = c(r)$ (for one orientation of e_{K_r}). We now deform g' on D_r rel $\dot{D}_r$, keeping the image of the deformation within the $4i$-skeleton, to a map g_r with the following properties:

(i) $q_r(D_r) \cap \overset{\circ}{e}_L = \emptyset$ for all 4i-links L other than K_r

(ii) Let $O(s,r)$ denote $q_r^{-1}(\overset{\circ}{e}_{K_r}) \cap S$ for all 4i-cells S with $r \subset \partial s$. Then $q_r|O(s,r) \to \overset{\circ}{e}_{K_r}$ is transverse to the "cone-point" of e_{K_r}, with multiplicity $-k_r$.

Point (ii) is understood in the following sense: an orientation for r has been implicitly chosen (so that $c(r) > 0$); this induces orientations on all s incident to r. As well, an orientation has been given for e_{K_r}, viz; the one which makes $k_r \ell_i(e_{K_r}) = c(r)$. It is with respect to these orientations that the degree of $q_r|O(s,r)$ is to be $-k_r$. That the proposed deformation to q_r can be made is a straightforward consequence of the fact that $\mathcal{G}_{n,k}$ is connected.

Since, on each D_r, the deformation has been constructed rel $\dot{D}_r$, it is clear that the union of all such deformations over all D_r with $c(r) \neq 0$ is extended in a trivial way to a deformation of g'. Denote the final stage of this deformation by g''.

It is now clear that $g'': M \to \mathcal{G}_{n,k}$ has the following property: for an arbitrary oriented 4i-cell s of S: $(g'')^{-1} \bigcup_L *_L \cap s$ is a finite collection of points (where $*_L$ now denotes the cone-point of e_L. Now for each such point p, we get a number $\lambda(p)$ defined by $\lambda(p) = \ell_i(e_p, O_r)$ (where e_p is the unique 4i-cell of $\mathcal{G}_{n,k}$ with $g''(p) \in e_p$) and O_r is the orientation induced by that of s near p. The important property is that $\sum_p \lambda(p) = 0$. (Note: with the original Gauss map $g(F)$ in place of g'' used to define $\lambda(p)$, we would have had $\sum\lambda(p) = \hat{\ell}_i(s)$).

Now, we simply proceed as follows: using the procedure specified in the proof of 4.2, working with the map g'' (which, perforce, is covered by a bundle map $TM \to \gamma_{n,k}$) we produce an immersion $f_1: M \to R^{n+k}$, which supports a triangulation T_1 with respect to which f_1 is a linear homeomorphism on each simplex. It further follows, from the method elaborated in the proof of 4.2 for constructing this complex that, given (n-4i)-dimensional simplex σ of T_1, whose formal link with respect to the immersion is L_σ, then V_{L_σ} is either "degenerate," in the sense that it coincides with some V_K where dim K < dim L or $V_{L_\sigma} = V_K$, where K is a formal link of dimension 4i such that $*_K$ = cone point of e_K is in the image of g". In the first case, as we have noted $\ell_i(e_L) = 0$. Therefore, given a 4i-cell s of S', in order to determine $\hat{\ell}_i(M,f_1)(s)$ we need merely count up the intersections of s with each σ in the latter category and multiply by the appropriate $\ell_i(e_{L_\sigma})$. [Here, we may assume that S has been isotoped to put it in general position with respect to T_1, and so that the intersection of s with an (n-4i)-simplex σ occurs at the barycenter of σ.]

Concentrating on σ of this kind, let us group them, i.e. to each such σ let $K(\sigma)$ be chosen such that $V_{L_\sigma} = V_{K(\sigma)}$, $*K_{(\sigma)} \in$ im g". More unambiguously, with the notation of 4.2 in mind, $\sigma \subset M_{K(\sigma)}$ and, in fact $f_1 : \sigma \to X_{K(\sigma)}$, where $X_{K(\sigma)}$ is the linear (n-4i)-plane of R^{n+k} defined in §2. Now, if we orient s and e_K for all 4i-dimensional K with $*_K \in$ im g", we obtain thereby an orientation of e_{L_σ}, for all σ with $K_\sigma = K$. We claim that the sum of the intersection number of s with all such σ is equal to the algebraic multiplicity μ_K of $(g")^{-1} *_K$. This is because we may isotop S so that under the immersion f_1, $s \cap M_K \to V_K$ hits X_K transversally with intersection number μ_K. From this it follows immediately, since $\ell_i(e_{L_\sigma}) = \ell_i(e_{K(\sigma)})$ that $\hat{\ell}_i(m,f_1)(\sigma) = \sum\lambda_p = 0$.

Thus the theorem is proved.

In conclusion, we note that Theorem 4.6 suggests the following purely differential-geometric conjecture: Suppose M^n is a smooth manifold with $L_i(M^n) = 0$. Given a smooth regular cell decomposition of M^n, does there exist a Riemannian metric on M^n such that $\int_s \omega_i = 0$ for all 4i-cells s, where ω_i is the closed form representing L_i in deRham cohomology corresponding to that particular metric?

5. Immersions equivariant with respect to orthogonal actions on R^{n+k}

In this chapter we shall study the problem of immersing an n-manifold M^n in R^{n+k}, while respecting certain geometric restrictions much like those in §4, where, in addition there is given a locally-smooth PL action of the finite group π on M^n, and an (n+k)-dimensional orthogonal representation of π, and where the immersion $M^n \to R^{n+k}$ is to be equivariant with respect to these π actions.

We first note that, since π has a given $O(n+k)$ representation, it will act on the PL Grassmannian $\mathcal{G}_{n,k}$ constructed in §2 above. We describe this action briefly; given an element $m \in O(n+k)$ and a j-dimensional formal link $L = (U_L, \Sigma_L)$, m acts on L to produce $m \cdot L$ by:

$$m \cdot L = (U_{m\cdot L}, \Sigma_{m\cdot L})$$

$U_{m\cdot L} = m \cdot U_L$ where the action on the right is the standard action of $O(n+k)$ on the standard Grassmannian $G_{j,n+k-j}$.

$\Sigma_{m\cdot L}$ is the (triangulated) sphere which is the image of Σ_L under the homeomorphism m. It is immediate that this serves to define a dimension-preserving $O(n+k)$-action on the set of formal links. Furthermore, it is a straightforward consequence of definitions and of linearity that if σ is a simplex of Σ_L, then $m\cdot(L_\sigma) = (m\cdot L)_{m\sigma}$ (where $m\sigma$ is the simplex of $\Sigma_{m\cdot L}$ which corresponds to σ). In other words, this $O(n+k)$ action preserves face relations among formal links.

It follows that we shall find the diagram

$$\begin{array}{ccc} c\,\Sigma_{L_\sigma} & \xrightarrow{h(L,\sigma)} & c\,\Sigma_L \\ {\scriptstyle m}\downarrow & & \downarrow{\scriptstyle m} \\ c\,\Sigma_{m\cdot L_{m\cdot\sigma}} & \xrightarrow{h(m\cdot L,m\cdot\sigma)} & c\,\Sigma_{m\cdot L} \end{array}$$

to be strictly commutative and to preserve obvious cone structures.

Now the cells e_L of $\mathcal{G}_{n,k}$ are to be thought of as the spaces $c\Sigma_L$ mod certain identifications on the boundary. By dint of the observations above, for $m \in O(n+k)$ $m:c\Sigma_L \to c\Sigma_{m\cdot L}$ is consistent with those identifications, i.e. m induces a homeomorphism of closed cells $\phi_{m,L}:e_L \to e_{m\cdot L}$ and this is consistent with face relations, i.e. $\phi_{m,L}|\, e_{L_\sigma} = \phi_{m,L_\sigma}$. Thus m produces a cellular automorphism $\phi_m = \bigcup \phi_{m,L} : \mathcal{G}_{n,k} \to \mathcal{G}_{n,k}$. It is trivially checked that the set of automorphisms $\{\phi_m\}_{m\in O(n,k)}$ yields a group action of $O(n+k)$ on $\mathcal{G}_{n,k}$ (with the discrete topology on $O(n+k)$).

We note, paranthetically, that the action of $O(n+k)$ on $\mathcal{G}_{n,k}$ extends to an action of the general linear group $GL(n+k;R)$. In this case $m\cdot L$ would have to be specified by letting $U_{m\cdot L} = m\cdot U_L$ (under the standard action of $GL(n+k);R)$ on $G_{j,n+k-j}$ and letting $\Sigma_{m\cdot L}$ be the <u>radial projection</u> of $m(\Sigma_L)$ on $S_{U_{m\cdot L}}$. The rest of the construction above then goes through essentially unchanged.

We wish, as well, to show that the PL bundle $\gamma_{n,k}$ is an $O(n+k)$-bundle (i.e. is acted upon by $O(n+k)$ with the discrete topology). Note first that for $m \in O(n+k)$, $V_{m\cdot L} = mV_L$ (where V_L, as before denotes the PL n-plane in R^{n+k} associated to the link L.) This follows since by definition, $V_L = Q_L \oplus X_L$, $V_{m\cdot L} = Q_{m\cdot L} \oplus X_{m\cdot L}$, and, while m is obviously a homeomorphism of Q_L onto $Q_{m\cdot L}$, since m is orthogonal, we must also have $mX_L = mU_L^{\perp} = (mU_L)^{\perp} = U_{m\cdot L}^{\perp} = X_{m\cdot L}$.

Moreover, recall the tautologica map $G: \mathcal{G}_{n,k} \to R^{n+k}$. Direct inspection of this construction reveals that G is an equivariant map $\mathcal{G}_{n,k} \to R^{n+k}$ with respect to the action of $O(n+k)$. Recall now the decomposition of $\mathcal{G}_{n,k}$ into subspaces $\bar{e}_L$, L a formal link. Clearly $\phi_m(\bar{e}_L) = \bar{e}_{m\cdot L}$, so that this decomposition is preserved under the action of $O(n+k)$. Finally, recall that over $\bar{e}_L$, $\gamma_{n,k}$ is defined locally as $(G|\bar{e}_L)^*TV_L$. Since

$$\begin{array}{ccc} \bar{e}_L & \xrightarrow{\phi m} & \bar{e}_{m\cdot L} \\ \downarrow G & & \downarrow G \\ V_L & \xrightarrow{m} & V_{m\cdot L} \end{array}$$

strictly commutes, there is a natural bundle map $\gamma_{n,k}|\bar{e}_L \to \gamma_{n,k}|\bar{e}_{m\cdot L}$ covering $\phi_m|\bar{e}_L \to \bar{e}_{m\cdot L}$, which is defined by the map on tangent bundles $TV_L \to TV_{m\cdot L}$ induces by the (PL) homeomorphism $m|V_L \to V_{m\cdot L}$. It remains to check that this bundle map associated to m is consistently defined over all of $\mathcal{G}_{n,k}$, i.e., that the union of such local bundle maps is itself a well-defined global bundle map $\gamma_{n,k} \to \gamma_{n,k}$ covering ϕ_m. To see this, we merely compare definitions on spaces of the form $\bar{e}_L \cap \bar{e}_K$. If this is not void, we must have $K < L$ (or $L < K$), so, assuming this, we have $G:(\bar{e}_L \cap \bar{e}_K) \to \text{int}\,(V_L \cap V_K)$ and thus TV_L and TV_K are naturally identified over the image under G of $\bar{e}_L \cap \bar{e}_K$. Hence the bundle G^*TV_L is naturally identified with G^*TV_K over $\bar{e}_L \cap \bar{e}_K$. Thus, given $m \in O(n+k)$ we have

$$\begin{array}{ccc} \bar{e}_L \cap \bar{e}_K & \xrightarrow{\phi_m} & \bar{e}_{m\cdot L} \cap \bar{e}_{m\cdot K} \\ \downarrow G & & \downarrow G \\ \text{int}(V_L \cap V_K) & \xrightarrow{m} & \text{int}(V_{m\cdot L} \cap V_{m\cdot K}) \end{array}$$

and so, over $\bar{e}_L \cap \bar{e}_K$ the map $\gamma_{n,k}|\bar{e}_L \cap \bar{e}_K \to \gamma_{n,k}|\bar{e}_{m\cdot L} \cap \bar{e}_{m\cdot K}$ may equally well be viewed as arising from $TV_L \to TV_{m\cdot L}$ (naturally covering the homeomorphism $m: V_L \to V_{m\cdot L}$) or from $TV_K \to TV_{m\cdot K}$, these two maps being identical over closed subsets of $\mathrm{int}(V_L \cap V_K)$. Hence, as stated, the cellular automorphism ϕ_m is naturally covered by a bundle map which we may call $\hat{\phi}_m: \gamma_{n,k} \to \gamma_{n,k}$. That this makes $\gamma_{n,k}$ into an $O(n+k)$-bundle (discrete topology on $O(n+k)$) is immediate.

Again, we digress briefly to consider the case of $GL(n+k;R)$. As we have seen, this larger group acts on $\mathcal{G}_{n,k}$ by cellular automorphisms. Note, however that the tautological map G is no longer equivariant with respect to this enlarged action, nor are the manifolds V_L even preserved. Nonetheless, it is still possible to view $\gamma_{n,k}$ as a $GL(n+k;R)$ bundle. Recall the alternative definition of $\gamma_{n,k}$ from §2 above. This involves for any L, considering $TV_L|b_L$ and then making identifications on the total space to cover those which collapse b_L onto $e_L \subset \mathcal{G}_{n,k}$. (Recall that b_L is merely the polyhedron whose vertices are those of $c\Sigma_L$ and whose simplices are the convex hulls in R^{n+k} of the simplex-spanning sets of vertices of $c\Sigma_L$. For the purposes of this construction, it is convenient to consider b_L, rather than the isomorphic $c\Sigma_L$, as the preimage of e_L.) For $m \in GL(n+k;R)$ we consider the induced map $b_L \to b_{m\cdot L}$. To cover this by a bundle map $TV_L|b_L \to TV_{m\cdot L}|b_{m\cdot L}$, we may proceed as follows: $TV_L = TQ_L \oplus X'_L$ where X'_L is the subbundle of TR^{n+k} of vectors parallel to X_L. The given map $m: b_L \to b_{m\cdot L}$ extends to a homeomorphism in $Q_L \to Q_{m\cdot L}$ in a completely obvious way, and thus $TQ_L|b_L \to TQ_{m\cdot L}|b_{m\cdot L}$ is well-defined. On the other hand, there is a bundle map $X'_L|b_L \to X'_{m\cdot L}|b_{m\cdot L}$ defined, fiberwise, by taking $X'_L(y) \to x'_{m\cdot L}(my)$ by $X'_L(y) \to m^*(X'_L(y)) \xrightarrow{\text{proj}} X'_{m\cdot L}(my)$. The Whitney sum of these two stipulated bundle maps is the desired bundle map $TV_L|b_L \to TV_{m\cdot L}|b_{m\cdot L}$. We claim that this family of bundle maps is once more compatible with face relations on formal links and thus obtain over $\phi_m: \mathcal{G}_{n,k} \to \mathcal{G}_{n,k}$ the bundle map $\hat{\phi}_m: \gamma_{n,k} \to \gamma_{n,k}$ and hence, the $GL(n+k; R)$ (discrete) structure on $\gamma_{n,k}$.

For the remainder of this section, however, we shall make no further use of the larger group $GL(n+k; R)$ and shall confine our attention to $O(n+k)$.

Therefore, if π is a finite group acting on the PL manifold M^n (locally smoothly in the sense of [Br] and, for some appropriate triangulation of M, simplicially), we wish to consider orthogonal representations $\pi \to O(n+k)$ and equivariant immersions $f: M^n \to R^{n+k}$. If a triangulation be chosen for M^n so that the π-action is simplicial and so that f is a linear embedding on each simplex, we look at the Gauss map $g(f): M_o^n \to \mathcal{G}_{n,k}$, along with its covering bundle map $\hat{g}(f): TM_o^n \to \gamma_{n,k}$. Note that for $\partial M^n \neq \emptyset$, M_o^n is a π-invariant subspace of M^n. We may thus state

<u>5.1 Lemma</u>. The Gauss map $\hat{g}(f), g(f): TM_o^n, M_o^n \to \gamma_{n,k}, \mathcal{G}_{n,k}$ is a map of π-bundles, where the π-structure on the space $\mathcal{G}_{n,k}$ and the bundle $\gamma_{n,k}$ is that induced from the $O(n+k)$-structure defined above by the given representation $\pi \to O(n+k)$.

This lemma needs no proof beyond some routine observations directly following from definitions, which we leave to the reader as an elementary exercise.

In seeking to generalize the immersion results of §4 to the equivariant case, it is clearly natural to consider subcomplexes H of $\mathcal{G}_{n,k}$ which are geometric, in the sense of Def. 4.1 and which, additionally, are invariant under the π-action on $\mathcal{G}_{n,k}$ associated to the representation $\pi \to O(n+k)$.

Correspondingly, there are restrictions on M^n. As stated above, we shall assume that the action of π on M^n is locally smooth. Bredon's definition [Br] is for the case of topological actions of compact groups, but with π-finite and acting simplicially on M^n, we shall stipulate that an action is taken to be locally smooth if and only if, for each orbit $\pi\cdot x$, $x \in M^n$, there exists an $O(n)$-representation of the isotropy group π_x, and a PL π-embedding $t\colon \pi \times_{\pi_x} D^n \to M^n$ taking $\pi \times_{\pi_x} D^n$ onto a closed π-tubular neighborhood of $\pi\cdot x$ with $t\pi \times_{\pi_x} \{0\} = \pi\cdot x$. Here, of course, D^n is given the π_x-structure of the representation.

Some useful facts follow from this notion of local smoothness. First, suppose θ is a subgroup of π such that π/θ is an orbit type for the action. Let M^θ be the set of points fixed by θ, and $M^{(\theta)} = \pi\cdot M^\theta$, while $M_\theta = \{x \in M \mid \pi_x = \theta\}$ and $M_{(\theta)} = \pi\cdot M_\theta$. Thus $M_\theta \subseteq M^\theta$, $M_{(\theta)} \subseteq M^{(\theta)}$. Local smoothness of the action implies that the π-components of $M_{(\theta)}$ are all PL submanifolds of M^n. Clearly, $M^{(\theta)} = \bigcup_{\Phi \supseteq \theta} M_{(\Phi)}$, and so $M^{(\theta)}$ is naturally stratified.

Consider a π-component P of $M_{(\theta)}$, and assume that is closed (topologically) in M^n, and let r denote its dimension as a manifold. P then has an open equivariant neighborhood. That is, there is a block - D^{n-r}-bundle ξ over P acted on by π which PL embeds equivariantly onto a closed neighborhood of P in M. The embedding is the identity on the 0-section of ξ.

A new condition which we now impose on M^n is the so-called <u>Bierstone condition</u> [Bi], suitably adopted to the PL case at hand. This condition, introduced by Bierstone in his study of the equivariant generalization of Gromov theory, stands in place of the usual "no closed components" condition of the non-equivariant Gromov (or Hirsch-Poenaru) results. We formulate this condition as follows:

Consider the set of π-components of $M^{(\theta)}$, θ ranging over all possible isotropy groups. This set is partially ordered by inclusion. Consider now an element minimal with respect to this ordering. Clearly, a space P of this description is a π-component of some

$M^{(\theta)}$, with $\pi_x = \theta$ for all $x \in P$, and thus, P is a π-component of $M_{(\theta)}$ as well, so that P is a manifold topologically closed in M^n.

The Bierstone Condition, then, may be stated: All such P are non-closed as manifolds. (I.e., each topological component of P is a handlebody with no top-dimensional handles.]

(In point of fact, Bierstone's original formulation of the condition is somewhat different; the characterization given above is subsequently shown to be equivalent).

Note that the Bierstone condition automatically guarantees that M^n itself is non-closed.

We may now state the main result of this section, an equivariant counterpart to 4.2.

5.2 Theorem. Let $\mathcal{H}$ be a geometric subcomplex of $\mathcal{G}_{n,k}$, invariant under π. Let M^n be a PL locally-smooth π-manifold satisfying the Bierstone condition.

Suppose $f: M^n \to \mathcal{H}$ is an equivariant map covered by a π-bundle map $\hat{f}: TM^n \to \gamma_{n,k}|\mathcal{H}$.

Then there is a π-equivariant immersion $f: M^n \to R^{n+k}$ so that the equivariant Gauss map $g(f): M^n \to \mathcal{G}_{n,k}$ has its image in $\mathcal{H}$.

Proof: As previously observed, the action of $O(n+k)$, and therefore of π, upon $\mathcal{G}_{n,k}$ preserves the decomposition of $\mathcal{G}_{n,k}$ into the subspaces $\bar{e}_L$, viz, $\phi_m \bar{e}_L = \bar{e}_{m \cdot L}$. Let $\bar{\mathcal{H}} = \bigcup_{e_L \subset \mathcal{H}} \bar{e}_L$; then $\bar{\mathcal{H}}$ contains $\mathcal{H}$ as a deformation retract and is itself π-invariant. As in the proof of 4.2, we wish to deform the map $f: M \to \mathcal{H} \subset \bar{\mathcal{H}}$, staying within $\bar{\mathcal{H}}$, so that the new map $f': M \to \bar{\mathcal{H}}$ has $M_L = f'^{-1}(\bar{e}_L)$ a codimension-0 submanifold of M^n. Moreover, we wish to have this deformation π-equivariant.

The analogous step in the non-equivariant case 4.2 was virtually trivial, based on general position considerations. However, with a group action present, further argument is needed because, as is well-known, the problem of equivariantly deforming an equivariant map so as to put it into general position with respect to some invariant subvariety of the target space may meet some non-trivial obstructions. Let us rephrase the problem somewhat:

Let $\mathcal{L} = L_1 < L_2 \ldots < L_r$ be a sequence of formal links linearly ordered by face relations, $e_{L_i} \subset \overline{\mathcal{H}}$. Then $\bar{e}_{\mathcal{L}} = \bigcap_{i=1}^{r} \bar{e}_{L_i}$ is a subspace of $\overline{\mathcal{H}}$ with a trivial r-1-disc bundle neighborhood. So, in fact, $\{\bar{e}_{\mathcal{L}}\}$ is a stratification of $\overline{\mathcal{H}}$, and we may therefore say that if a map, e.g. $f: M \to \overline{\mathcal{H}}$ is transverse to $\{\bar{e}_{\mathcal{L}}\}$, then $\{f^{-1}\bar{e}_{\mathcal{L}}\}$ will be a stratification of M, with $f^{-1}\bar{e}_{\mathcal{L}}$ a submanifold of codimension $r-1$, $r = \#\mathcal{L}$. Thus, we shall show that the given map f may be equivariantly deformed to $f': M \to \overline{\mathcal{H}}$, f' of this type.

The key observation is the following general principle: Suppose Q is acted on by the finite group Γ, and Q is, furthermore, provided with a stratification invariant under Γ. As we have noted above, the problem of equivariantly deforming an equivariant map $\phi: W \to Q$, (W a Γ-submanifold to a transverse map is, in general nontrivial. However, the following result gives a sufficient condition on the Γ-space Q for obstructions to vanish.

5.3 Proposition. If the orbit space Q/Γ is stratified so that the projection map $Q \to Q/\Gamma$ is transverse to the stratification, then any equivariant map $\phi: W \to Q$ may be equivariantly deformed to ϕ' so that $W \xrightarrow{\phi} Q \to Q/\Gamma$ is transverse to the stratification; thus ϕ' itself is transverse to the stratification on Q induced by projection.

The proof of this proposition is routine, and in any case, follows from the general theory of obstructions to equivariant transversality.

To return to the case at hand let $p\mathcal{L} = p\cdot L_1 < pL_2 \ldots < p\cdot L_r$ where $\mathcal{L} = L_1 < L_2 \ldots < L_r$. Define $\bar{e}_{\pi\mathcal{L}} \subset \bar{\mathcal{H}}/\pi$ to be the image of $\bar{e}_{\mathcal{L}}$ (or of $\bar{e}_{p\mathcal{L}}$, $p \in \pi$) under projection. We claim that $\bar{e}_{\pi\mathcal{L}}$ has a tubular neighborhood of the form $\bar{e}_{\pi\mathcal{L}} \times D^{r-1}$ in $\bar{\mathcal{H}}/\pi$. So, in particular $\{\bar{e}_{\pi\mathcal{L}}\}$ (indexed by all π-orbits of multi-indices) is a stratification of $\bar{\mathcal{H}}/\pi$. Moreover,

<u>5.4 Lemma</u>. The projection $\bar{\mathcal{H}} \to \bar{\mathcal{H}}/\pi$ is transverse to the stratification $\{\bar{e}_{\pi\mathcal{L}}\}$; therefore the map $f\colon M^n \to \bar{\mathcal{H}}$ is equivariantly deformable to f', with f' transverse to $\{\bar{e}_{\mathcal{L}}\}$.

5.4 follows by direct inspection of the constructions for $\{\bar{e}_L\}$, and of the definition of the action of π on $\mathcal{G}_{n,k}$.

We now may relabel f' as f, with $M_{\mathcal{L}} = f^{-1}\bar{e}_{\mathcal{L}}$ now a codimension(r-1)-submanifold of M for all $\mathcal{L}$, ($r = \#\mathcal{L}$).

We we have, in particular $M_L = f^{-1}\bar{e}_L$ a codimension-0 submanifold of M.

Set $M_{(L)} = \bigcup_{p\in\pi} M_{p\cdot L}$, which is an invariant codimension-0 submanifold of M.

We may assume that the new f is covered by a π-bundle map $TM^n \xrightarrow{\hat{f}} \gamma_{n,k} | \bar{\mathcal{H}}$.

Let $V_{(L)} = \coprod_{p\in\pi} V_{p\cdot L}$, where $\coprod$ denotes abstract disjoint union. Consider the map $G \circ f\colon M \to R^{n+k}$. Since $M_L \cap M_{p\cdot L} = \emptyset$ unless $L = p\cdot L$, we see that $G \circ f|M_{(L)}$ is factored uniquely as an equivariant map $f_{(L)}\colon M_{(L)} \to V_{(L)} \to R^{n+k}$, where $2_{(L)}\colon V_{(L)} \to R^{n+k}$ is merely the obvious inclusion on each $V_{p\cdot L} \subseteq V_{(L)}$.

As well, the bundle map $\hat{f}|TM_{(L)} \to \gamma_{n,k}$ determines an equivariant bundle map $\hat{f}_{(L)}\colon TM_{(L)} \to TV_{(L)}$ since on $\bar{e}_{\pi L} = \bigcup_{p\in\pi} \bar{e}_{p\cdot L}$, $\gamma_{n,k}$ is, essentially, the pullback of $TV_{(L)}$.

The idea of the proof is to turn each map $M_{(L)} \to V_{(L)}$ into an equivariant codimension-0 immersion. The tool we use is Bierstone's generalization of the Phillips-Gromov-Hirsch theory restricted to the immersion problem and translated into the PL category. We shall state the result we actually need as a lemma. First, however, we extend the Bierstone condition to cover the relative case.

Let N^n be a locally-smooth π-manifold and W^n an invariant codimension-0 submanifold topologically closed in N^n, so that $Z^n = \overline{M-W}$ is also an invariant codimension-0 submanifold. Let $Z_j^{\theta_i}$ be the minimal elements of the stratification of Z arising from the π-action, θ_i varying over the isotropy subgroups of π. We shall say that the pair M^n, W^n satisfies the <u>relative Bierstone condition</u> if and only if each connected component K of each $Z_j^{\theta_i}$ can be obtained from a collar on $K \cap \partial W$ by a sequence of handle-attachments not involving top-dimensional handles.

<u>5.5 Lemma</u>. Let M^n, $W^n \subset M^n$ be a pair of locally-smooth π-manifolds satisfying the relative Bierstone conditions. Let $f: M^n \to V^n$ be an equivariant map, V^n a π-manifold, and assume that $f|W^n$ is an immersion. Moreover, let f be covered by a bundle map of π-bundles $\hat{f}: TM^n \to TV^n$, so that, over W^n, this coincides with the bundle map induced by the immersion.

Then f may be equivariantly deformed, rel. W^n, to an equivariant immersion h. Moreover, the deformation from f to h may be covered by a deformation through π-bundle maps of $\hat{f}$ to the bundle map induced by h.

We shall not attempt a proof of this lemma here. We note that it arises from the fact that the Bierstone theorem [Bi] may certainly be reproved from scratch in the much more limited context of smooth equivariant immersions. That is, Bierstone's techniques, a fortiori, work as well in proving an equivariant version of the Hirsch immersion theorem as they do in proving an equivariant Gromov theorem. But then, they may be combined with the techniques used by Hirsch and Poenaru [H-P] to extend the smooth Hirsch immersion theorem [Hi] to the PL case. The stated lemma is the result of carrying through this program. This is also studied in [Mi].

Turning back, then, to the proof of 5.2 we shall carry through an inductive argument on the dimension of the formal links involved. Recall the map $M_{(L)} \to V_{(L)}$ through which $G \circ f|M_{(L)}$ factors, and the bundle map $TM_{(L)} \to TV_{(L)}$ (both equivariant). For the sake of establishing terminology, let A^n denote an arbitrary locally-smooth PL π-manifold. Recall that each orbit π_x, $x \in A$, has an equivariant tubular neighborhood of the form $\pi \times_{\pi x} D^n$, for some n-dimensional representation of π_x. A codimension-0 submanifold $A_1^n \subset A^n$ shall be called a punctured A^n if it consists of A^n with the interior of such an equivariant tubular neighborhood of an orbit removed. A_1^n shall be called a multiply punctured A^n if it consists of A^n with the (disjoint) interiors of several such neighborhoods of orbits removed.

We may now state our inductive hypothesis.

Hypothesis $H(j)$: (a) There are π-equivariant maps $h_{(L)}: M_{(L)} \to V_{(L)}$ covered by π-bundle maps $\hat{h}_{(L)}: TM_{(L)} \to TV_{(L)}$. Moreover on each $M_{(L)} \cap M_{(K)}$, $\iota_{(L)} \circ h_{(L)} = \iota_{(K)} \circ h_{(K)}$ where $\iota_{(L)}, \iota_{(K)}$ are the obvious maps $V_{(L)}, V_{(K)} \to R^{n+k}$.

(b) For $\dim L < j$, $h_{(L)}$ is an immersion on $M^-_{(L)}$, a multiply punctured $M_{(L)}$, and the bundle map $\hat{h}_{(L)}|M^-_{(L)}$ is merely that defined by the immersion.

(c) The map $h: M^n \to R^{n+k}$ defined by $h(x) = \iota_{(L)} \circ h_{(L)}(x)$, $x \in M_{(L)}$, is an immersion on

$$M^{(j-1)} = \bigcup_{\dim L<j} M^-_{(L)}.$$

To begin with, we obviously have $H(0)$ satisfied, with $f_{(L)}$, $\hat{f}_{(L)}$ as defined above playing the role of $h_{(L)}$, $\hat{h}_{(L)}$ (because clauses (b), (c) are vacuous for $j = 0$). We must therefore show that $H(j-1)$ implies $H(j)$. Assuming $H(j-1)$, consider $V_{(L)}$ for a typical $(j-1)$ link L. Without loss of generality, we may assume that $h(L)$ is an immersion on an invariant collar neighborhood $C^n_{(L)}$ of $M^{(j-2)} \cap M_{(L)}$. Note that $M^{(j-2)} \cap M_{(L)} = \bigcup_{\substack{p \in \pi \\ K < p \cdot L}} M_{p \cdot L} \cap M_K$.

We wish to extend this to an equivariant immersion $M_{(L)} \to V_{(L)}$, obtained by equivariently deforming $h_{(L)}$ rel $C_{(L)}$. The problem, though, is that $M_{(L)}, C_{(L)}$ does not necessarily satisfy the relative Bierstone condition. Therefore, let $A_{(L)} = \overline{M_{(L)} - C_{(L)}}$ and let $A_{(L),i}$ denote for some indexing set $\mathcal{J}$, with $i \in \mathcal{J}$, the various minimal elements of the stratification of $A_{(L)}$ associated with the action of π, such that $A_{(L),i}$ cannot be obtained from a collar on $A_{(L),i} \cap C_n$ by handle attaching without the use of top-dimensional handles.

[I.e. $A_{(L),i}$ will be a π-component of some $A^{(\theta)}_{(L)}$, $\theta < \pi$; $A_{(L),i}$ contains no π-component of $A^{(\phi)}$ for any $\phi > \theta$; thus $A_{(L),i}$ will be a manifold of some dimension, say r, and obtaining $A_{(L),i}$ from $A_{(L),i}$ $C_{(L)}$ involves r-handles.]

Now, for each such $A_{(L),i}$, pick some orbit, πx_i, $x_i \in A_{(L),i}$, and remove from $A_{(L)}$ a small equivariant tubular neighborhood. Clearly, such tubular neighborhoods may be assumed mutually disjoint. The removal of these neighborhoods yields a multiply-punctured $A_{(L)}$ and hence a multiply-punctured $M_{(L)}$ which we denote $M^-_{(L)}$. $M^-_{(L)}, C_{(L)}$ satisfies the relative Bierstone condition, hence we may deform $h_{(L)}$ rel $C_{(L)}$ to a map $h'_{(L)}$ which is an immersion on $M^-_{(L)}$. Covering this deformation by a deformation of $\hat{h}_{(L)}$ to $\hat{h}'_{(L)}$: $TM_{(L)} \to TV_{(L)}$, we obtain a π-bundle map which, over $M^-_{(L)}$ is that induced by the immersion $h'_{(L)}|M^-_{(L)} \to V_{(L)}$.

Moreover, we claim that the deformation of $h_{(L)}$ to $h'_{(L)}$ may be chosen so as to keep $M_{(L)} \cap M_{(K)}$ in $V_{(L)} \cap V_{(K)}$ (regarding the latter as a subspace of $V_{(L)}$) for all K with $L < K$. The argument here essentially replicates the analogous one in the proof of 4.2, the presence of a group action in this case notwithstanding. We therefore omit details.

Thus, if we perform the deformation of $h_{(L)}$ to $h'_{(L)}$ for each orbit (L), it is immediately apparent that $h_{(K)}$ (and $\hat{h}_{(K)}$ as well) may be suitably deformed (to $h'_{(K)}$, $\hat{h}'_{(K)}$ respectively) for all K of dimension $> j$, so that $\iota_{(K_1)} \circ h'_{(K_1)} = \iota_{(K_2)} \circ h'_{(K_2)}$ on $M_{(K_1)} \cap M_{(K_2)}$, K_1, K_2 of arbitrary dimension. Here, it is understood that $h'_{(K)} = h_{(K)}$ for $\dim K < j-1$. Moreover, it is understood that the deformations of $h_{(K)}$ to $h'_{(K)}$ satisfy the same compatibility conditions as the $h_{(K)}$ or $h'_{(K)}$ themselves. (The deformation is the trivial one for $\dim K < j-1$.)

Therefore we have defined, globally, a deformation of h to h', rel $M^{(j-2)}$. It is clear that h', defined locally as $\iota_{(K)} \circ h'_{(K)}$, is an immersion on $M^{(j-1)} = M^{(j-2)} \cup \bigcup_{(L)} M^{-}_{(L)}$. This competes our inductive step, obtaining hypothesis $H(j)$ from $H(j-1)$. Since $H(0)$ was known, we may assert $H(n+1)$. In consequence, we may assert the existence of a map $h\colon M^n \to R^{n+k}$, defined locally by $\iota_{(L)} \circ h_{(L)}$, which, moreover, is an immersion on $M^{(n)} \subset M$. We shall immerse all of M, or rather, a π-homeomorphic copy, as follows. Clearly $M^{(n)}$ is a multiply-punctured M, that is, $M^{(n)}$ has been obtained by removing some tubular neighborhoods of π-orbits.

For the sake of brevity, let us assume that M^n is compact with non-void boundary (the case when M^n is open is similar). Since the Bierstone condition holds on M^n, it follows that any π-component P of any $M_{(\theta)}$ must satisfy $P \cap \partial M^n \neq \emptyset$. (See the analogous Theorem 3.1 in [Bi]. Moreover, it is readily seen that $P^* = P/\pi$ is itself a manifold-with-boundary. Now suppose one of the orbit-neighborhoods removed to obtain $M^{(n)}$ from M is the neighborhood T of the orbit $\pi\cdot x$. $\pi\cdot x$ lies in one of the π-components P described above and the image of $T \cap P$ in P^* is a disc D^k,

$(k = \dim P)$ about $y = \text{image } \pi \cdot x \in P^*$. Now $\partial P^* = P^* \cap \partial M/\pi \neq 0$, so we pick an arc ρ from D^k to ∂P^* whose interior stays in int P^*. The inverse image of ρ in P is a disjoint family $\tilde{\rho}$ of paths from $\partial T \cap P$ to ∂P. We then excise from $M^{(n)}$ an equivariant tubular neighborhood of $\tilde{\rho}$. Doing this procedure successively for each orbit $\pi \cdot x$, (whose neighborhood has previously been removed to form $M^{(n)}$), we obtain at last the π-manifold $M_0 \subset M^{(n)} \subset M$. We claim, in fact, that M_0 is π-homeomorphic to M.

We show this under the simplifying assumption that obtaining $M^{(n)}$ from M involved the deletion of just one orbit-neighborhood. No essential loss of generality is created.

This assumption understood, let us note that M_0 may be viewed as M with the equivariant neighborhood of a π-invariant family of arcs deleted. (I.e., extend ρ in P^* by the "diameter" of the disc D^k, to produce σ. If $\tilde{\sigma}$ is the family of arcs covering σ, M_0 is M less an open π-neighborhood of $\tilde{\sigma}$-(inside endpoints). But then, let S be a product neighborhood $S = \sigma \times D^{k-1}$ of σ in P^*. Recalling that $P \subset M_\theta$, there will be a θ-bundle ξ with fiber D^{n-k} (a block bundle, more precisely) over S, such that $B = \pi \times_\theta \xi$ is π-isomorphic to the equivariant tube about $\tilde{\sigma}$ (whose interior was excised to produce M_0. Thus M may be decomposed as $M_0 \cup_{\dot{B}} B$ where $\dot{B} = \pi \times_\theta \dot{\xi} \cup \pi \times_\theta (\xi |$ endpoints $\tilde{\sigma})$. Now it is easy to find an explicit π-homeomorphism $M_0 \longleftrightarrow M$ so that $M \xrightarrow{\equiv} M_0 \subset M$ is π-homotopic to the identity.

Thus, changing notation, we regard $h|M_0 \to R^{n+k}$ as an immersion $h: M \to R^{n+k}$. Taking a π-equivariant triangulation of M so that h is linear embedding on simplices, we obtain an equivariant Gauss map $g(h)$. That $\text{im } g(h) \subseteq \mathcal{H}$ follows by the geometricity of $\mathcal{H}$, just as in the proof of 4.2. This completes the proof of 5.2, the main result of this section.

6. Immersions into triangulated manifolds (with R. Mladineo)

This chapter, and the succeeding one as well, are based on the doctoral thesis of my student, Regina Mladineo [Ml]. The idea of this work is to extend the immersion results of §4 (and their equivariant extensions of §5) to the case of immersions into an arbitrary triangulated manifold. Triangulated manifolds, it should be understood, are geometric, rather than merely topological objects. Roughly speaking, a particular combinatorial triangulation of a manifold stands, in relation to the underlying PL structure, as a Riemannian manifold to its underlying differentiable structure. In particular, by putting the standard metric on each simplex, such notions as PL curvature, in the sense of D. Stone, become defined [St_3],[St_4].

Given a triangulated manifold of dimension, it seems natural to ask whether it supports an "associated Grassmann bundle." Recall, by way of background, that any vector bundle ξ of fiber dimension $n+k$ has an associated Grassmann bundle $G_{n,k}(\xi)$ wherein the fiber $G_{n,k}(x)$ over any point x in the base space of ξ consists of the space of linear n-planes in the fiber vector space $\xi(x)$ over x. $G_{n,k}(\xi)$ supports a natural n-victor bundle $\alpha_{n,k}(\xi)$ whose fiber over the point in $G_{n,k}(\xi)$ representing the n-plane $Y \subset \xi_x$ is Y itself. Recall that Hirsch's immersion theorem thus may be restated, in terms of this construction: Let M^n, W^{n+k} be smooth manifolds and T_M, T_W their respective tangent vector bundles. Then according to Hirsch, $f: M^n \to W^{n+k}$ is homotopic to an immersion if and only if f lifts to $g: M^n \to G_{n,k}(T_W)$ so that $g^*\alpha_{n,k}(T_W) = T_M$. (Of course, we must assume [as usual] that M^n has no top-dimensional cell if $k = 0$.)

Moreover, pushing this result in the direction of the Gromov-Phillips theorem, we may assert that if $U \subset G_{n,k}(T_W)$ is some open set, and if f lifts to g with $\operatorname{im} g \subseteq U$, then f deforms to an immersion i with $\operatorname{im} h_i \subset U$, where h_i denotes the natural lift from W to $G_{n,k}(W)$ determined by the immersion.

Our aim in this section is to demonstrate an analog to this result. The first task, therefore, will be to construct, at least for a triangulated manifold W^{n+k}, the analogue to the Grassman bundle and its canonical n-plane bundle. This spece will be denoted $\mathcal{G}_{n,k}(W)$, and the canonical bundle by $\gamma_{n,k}(W)$. We will then study immersions $f: M^n \to W^{n+k}$, with M^n a PL manifold such that

(1) f is in general position with respect to the triangulation of W

(2) M^n is triangulated so that each star is embedded in W with each simplex linearly embedded in some simplex of W.

For such an immersion there is a natural Gauss map, i.e. a map $g: M^n \to \mathcal{G}_{n,k}(W)$ covered by a natural bundle map $TM^n \to \gamma_{n,k}(W)$. Moreover, geometric conditions on such an immersion f may be interpreted as a requirement that the Gauss map g has its image within a corresponding sub-complex $X \subset \mathcal{G}_{n,k}(W)$. The principal result of this section is that if X is a suitable subcomplex of $\mathcal{G}_{n,k}(W)$, that is, to extend our previous terminology, a <u>geometric</u> subcomplex, then any bundle map M^n, $TM^n \to X$, $\gamma_{n,k}(W)|X$ implies that there is an immersion $f: M^n \to W$ whose Gauss map g has image contained in X.

We now proceed to the construction of $\mathcal{G}_{n,k}(W)$.

$\mathcal{G}_{n,k}(W)$ is to be the union, mod certain identifications, of complexes $\mathcal{G}_{j,k}(\alpha^{j+k})$, where α^{j+k} is a j+k-simplex of W, and where $\mathcal{G}_{j,k}(\alpha^{j+k})$ is a copy of the PL Grassmannian $\mathcal{G}_{j,k}$ as defined in §2 above. It is understood, of course, that for $j < 0$, $\mathcal{G}_{j,k} = \emptyset$. The identifications in question are modeled on the simplicial complex W; that is, if $\alpha^{j+k} < \beta^{i+k}$ are simplices of W, we shall identify $\mathcal{G}_{j,k}(\alpha^{j+k})$ with a certain subcomplex of $\mathcal{G}_{i,k}(\beta^{i+k})$. This is done as follows:

For each dimension $j+k \quad n+k$ we choose, once and for all, a fixed isometric copy of the standard simplex Δ^{j+k} embedded linearly in R^{j+k} with its barycenter at the origin. (The choice of the particular isometric embedding $\Delta^{j+k} \subset R^{n+k}$ will not affect the definition of $\mathcal{G}_{n,k}(W)$.) For each simplex α^{j+k} of W pick an isometry that is, in effect, a 1-1 correspondence of vertices, $\phi_\alpha : \alpha^{j+k} \cong \Delta^{j+k} \subset R^{j+k}$.

It is then clear that if $\alpha^{j+k} < \beta^{i+k} \subset W$, there is a well-defined inclusion $t_{\alpha,\beta} : R^{j+k} \to R^{i+k}$ embedding R^{j+k} as a linear subspace of R^{i+k}. That is, the composition

$$\Delta^{j+k} \xrightarrow{\phi_\alpha^{-1}} \alpha < \beta \xrightarrow{\phi_\beta} \Delta^{i+k} \subset R^{i+k}$$

extends uniquely to an affine embedding $R^{j+k} \subset R^{i+k}$. If we then shift this embedding by the vector $-b$, where b is the barycenter of $\phi_\beta(\alpha)$, we obtain $t_{\alpha\,\beta}$. Furthermore, the map $t_{\alpha\beta}$ induces an inclusion $\iota_{\alpha,\beta} : \mathcal{G}_{j,k} \subset \mathcal{G}_{i,k}$. To see this, let $L = (U_L, \Sigma_L)$ be a formal link of dimension $(j,k;r)$. Let $\iota_{\alpha,\beta}(L) = L'$ be the $(i,k;r)$-link determined by $U_{L'} = t_{\alpha,\beta}(U_L)$ and $\Sigma_{L'}$ = the admissibly triangulated sphere $t_{\alpha,\beta}(\Sigma_{L'})$. Then we may think of $\iota_{\alpha,\beta}$ as defining an isomorphism $e_L \cong e_{L'}$. This is clearly compatible with face relations, and may be taken to specify an inclusion map $\iota_{\alpha,\beta} : \mathcal{G}_{j,k} \to \mathcal{G}_{i,k}$.

The definition of $\quad_{n,k}(W)$ is then immediate:

<u>6.1 Definition.</u> $\mathcal{G}_{n,k}(W)$ is the complex $\bigcup_{j,\alpha} \mathcal{G}_{j,k}(\alpha^{j+k})$ where $\mathcal{G}_{j,k}(\alpha^{j+k})$ is a copy of $\mathcal{G}_{j,k}$ and where $\mathcal{G}_{j,k}(\alpha^{j+k})$ is identified with its image $\iota_{\alpha,\beta}\,\mathcal{G}_{j,k}(\alpha) \subset \mathcal{G}_{i,k}(\beta)$ for each face relation $\alpha^{j+k} < \beta^{i+k}$.

It must be noted, of course, that if $\alpha < \beta < \gamma$, then $\iota_{\alpha,\gamma} = \iota_{\gamma,\beta} \circ \iota_{\alpha,\beta}$.

In generalizing $\mathcal{G}_{n,k}$ to $\mathcal{G}_{n,k}(W)$ we also find it convenient to generalize the canonical map $\mathcal{G}_{n,k} \to R^{n+k}$ to a canonical map $G_W: \mathcal{G}_{n,k}(W) \to W$. Let α^{j+k} be a simplex of W and $\mathcal{G}_{j,k}(\alpha) \subset \mathcal{G}_{j,k}(W)$ the corresponding copy of $\mathcal{G}_{j,k}$. We set $V^{\alpha}_{L_\alpha} = \phi_{\alpha}^{-1}(V_L \cap \Delta^{j+k})$ for L a $(j,k;r)$ formal link. Let $\alpha^{j+k} < \beta^{i+k}$ and let $V^{\alpha\beta} = \phi_{\beta}^{-1}(V_{\iota_{\alpha,\beta}(L)} \cap \Delta^{i+k})$. Thus $V^{\alpha\beta}_L$ is a $j+k$ - disc in β, with $V^{\alpha\beta}_L \cap \alpha = V^{\alpha}_L$. Now, let $\bar{V}^{\alpha}_L = \bigcup_{\alpha<\beta} V^{\alpha\beta}_L$. It is seen then that $\bar{V}^{\alpha}_L$ is a PL n-disc embedded in $st(\alpha,W)$.

With these constructions in hand, we proceed to the definition of G_W.

Let $\hat{S}^{j+k-1}$ denote the sphere in R^{j+k} centered at the origin and circumscribed by the chosen copy of Δ^{j+k}, i.e. $\hat{S}^{j+k-1}$ is contained in Δ^{j+k} and tangent to the maximal proper faces of Δ^{j+k} at their respective berycenters. Recall that $\mathcal{G}_{j,k}$ has a simplicial subdivision where, for any $(j,k;r)$ link L, the cell e_L is a subcomplex whose simplices are the images of the simplices of the complex (see §2, p.2.9). Thus, we may simplicially triangulate $\mathcal{G}_{n,k}(W)$ by letting the simplices be those of each $\mathcal{G}_{j,k}(\alpha^{j+k})$. This gives a simplicial structure to $\mathcal{G}_{n,k}(W)$ since $\iota_{\alpha,\beta}\mathcal{G}_{j,k}(\alpha) \subset \mathcal{G}_{i,k}(\beta)$ is a simplicial inclusion.

Hence, to specify the canonical map $G_W: \mathcal{G}_{n,k}(W) \to W$, it will suffice to define it on vertices, provided $G_W(v) \in \alpha$ whenever $v \in \mathcal{G}_{j,k}(\alpha)$, for in this case convex linear extension will be well-defined.

This in mind, let v be a vertex of $\mathcal{G}_{j,k}(\alpha^{j+k})$, and let α, moreover, be the smallest simplex for which this holds. Thus, v corresponds to some vertex v_0 of the "standard" $\mathcal{G}_{j,k}$. Recall the tautological map $G: \mathcal{G}_{j,k} \to R^{j+k}$. Let $p(x)$ denote radial projection on $\hat{S}^{j+k-1}$ if $x \neq 0$, 0 if $x = 0$. So, in particular,

$pG(v) \in \Delta^{j+k}$. Let $G_W(v) = \phi_\alpha^{-1} pG(v)$.

6.2 Definition. $G_W : \mathcal{G}_{n,k}(W) \to W$ is the convex linear extension of G_W as already defined on the vertices of $\mathcal{G}_{n,k}(W)$.

With the definition of G_W in mind, recall the definition of the space $\bar{e}_L \subset \mathcal{G}_{j,k}$ for L a formal $(j,k;r)$-link. Let α^{j+k} be a simplex of W and L a $(j,k;r)$-link such that $e_L \subset \mathcal{G}_{j,k}(\alpha)$ but $e_L \neq \iota_{\gamma,\alpha} e_{L'}$ for any $\gamma < \alpha$ (i.e. L is not $\iota_{\gamma,\alpha} L'$ for any pair γ, L' where L' is an $(h,k;r)$ link and $\gamma^{h+k} < \alpha^{j+h}$). Thus we have $\bar{e}^{\beta}_{\iota_{\alpha\beta}(L)} \subset \mathcal{G}_{i,k}(\beta^{i+k})$ for all $\beta > \alpha$. Let $\tilde{e}^{\alpha}_L = \bigcup_{\alpha<\beta} \bar{e}^{\beta}_{\iota_{\alpha,\beta}(L)} \subset \mathcal{G}_{n,k}(W)$.

We make the following observation leaving the elementary task of verification to the reader:

6.3 Proposition. $G_W(\tilde{e}^{\alpha}_L) \subset \bar{V}^{\alpha}_L \subset W$.

We may now specify the n-dimensional PL bundle $\gamma_{n,k}(W)$ locally by specifying that $\gamma_{n,k}(W)|\tilde{e}^{\alpha} = (G_W|\tilde{e}^{\alpha}_L)^* T\bar{V}^{\alpha}_L$.

Noving that the various spaces $\tilde{e}^{\alpha}_L$ cover $\mathcal{G}_{n,k}(W)$ we need merely show local consistency of these various local definitions to obtain a global PL n-bundle $\gamma_{n,k}(W)$. To this end, we note that $\tilde{e}^{\alpha}_L \cap \tilde{e}^{\beta}_K \neq \emptyset$ only if $\alpha < \beta$ (or $\beta < \alpha$) and $\iota_{\alpha,\beta} L < K$ (or $\iota_{\beta,\alpha} K < L$). In this case $\bar{V}^{\beta}_K$ is a codimension-0 submanifold of $\bar{V}^{\alpha}_L$ and it is thus clear that the two local definitions $(G_W|\tilde{e}^{\alpha}_L)^* T\bar{V}^{\alpha}_L$ and $(G_W|\tilde{e}^{\beta}_K)^* T\bar{V}^{\beta}_K$ agree on $\tilde{e}^{\alpha}_L \cap \tilde{e}^{\beta}_K$ through a canonical identification.

Having defined $\mathcal{G}_{n,k}(W)$ and, with the aid of the tautological map G_W, the bundle $\gamma_{n,k}(W)$, it remains for us to sec how a Gauss map M^n, $TM^n \to \mathcal{G}_{n,k}(W)$, $\gamma_{n,k}(W)$ arises from a generic PL immersion $M^n \to W$. In particular, we shall assume that the immersion f is in general position with respect to each simplex of W. Thus if

α is a $(j+k)$-simplex of W, $M_\alpha = f^{-1}\alpha$ is a j-dimensional submanifold of M with boundary $\partial M_\alpha = f^{-1}(\partial\alpha)$. We shall assume henceforth that M is triangulated so that $f^{-1}\alpha$ is a subcomplex. Let $M_\alpha^o \subset M_\alpha$ be defined by $M_\alpha^o = \bigcup_{\tau \not\subset \partial M_\alpha} \tau_\alpha^*$ where τ is a simplex of M_α (not contained in the boundary) τ_α^* its dual cell (in the manifold M_α). As usual, $M_\alpha^o = M_\alpha$ - (collar on bdy.). On M_α^o, the Gauss map g is to be defined as follows: $\phi_\alpha \circ f$ immerses M_α in R^{j+k}, yielding a Gauss map $g_\alpha : M_\alpha^o \to \mathcal{G}_{j,k}$. Identifying the latter space with $\mathcal{G}_{j,k}(\alpha) \subset \mathcal{G}_{n,k}(W)$ we get $g|M_\alpha^o \to \mathcal{G}_{n,k}(W)$. We now wish to extend this definition to all of M (or to $M^o = M$ - (collar on ∂M) if $\partial M \neq \emptyset$.)

Let $\tilde{\sigma}$ be a simplex of M_γ, $\tilde{\sigma} \not\subset \partial M$ where γ is of maximal dimension $n+k$ in W. Let $\Delta = \Delta^{n+k} = \phi_\gamma(\gamma)$, and set $\sigma = \phi_\gamma \circ f(\tilde{\sigma})$, $\lambda(\tilde{\sigma},\gamma) = \phi_\gamma(f(st(\tilde{\sigma},M)) \cap \gamma)$, $\lambda_\beta(\tilde{\sigma},\gamma) = \beta \cap \lambda(\tilde{\sigma},\gamma)$ where β is a face of Δ having $\sigma \subset \beta$.

We now construct a certain complex K (depending on $\tilde{\sigma},\gamma$). Given $\beta < \Delta$ let $\bar{\beta}$ denote the complementary face of Δ, i.e. $\Delta = \beta * \bar{\beta}$. Let $\sigma(\bar{\beta})$ denote the reflection (in R^{n+k}) of $\bar{\beta}$ through b_σ, i.e.

$$\sigma(\bar{\beta}) = \{y = b_\sigma - (x - b_\sigma) \mid x \in \bar{\beta}\}$$

Now let α designate the smallest face of Δ such that $\sigma \subset \alpha$. Finally, set

$$K = \lambda(\tilde{\sigma},\gamma) \cup \bigcup_{\alpha \leq \beta < \gamma} (\sigma(\bar{\beta}) * \lambda_\beta(\sigma,\gamma))$$

Here, it is understood that the terms in this union which take the form of joins are naturally embedded in R^{n+k}, as is the complex K. Note that K is an n-manifold (with boundary) containing σ as a simplex and therefore, we have the formal link $L(\sigma,K)$. Correspondingly $e_{L(\sigma,K)}$ may be thought of as a cell of $\mathcal{G}_{n,k}(\gamma)$.

Recall that for each $\delta < \gamma$, we have already defined $\tilde{g}_\delta|\tilde{\sigma}^* \cap M_\delta^o$, and im $\tilde{g}_\delta \subset \mathcal{G}_{n,k}(\gamma) \subset \mathcal{G}_{n,k}(W)$. It is obvious by inspection that im $\tilde{g}_\delta$ lies, in fact, in the cell $e_{L(\sigma,K)}$ with im $\tilde{g}_\delta$ moreover in the boundary of this cell. We define g on $\tilde{\sigma}^* \cap \gamma$ by linear extension; that is, if we let $\tilde{\sigma}_0^*$ be that part of $\tilde{\sigma}^* \cap \gamma$ on which g has already been described by the various $\tilde{g}_\delta$, then this map factors through $\Sigma_{L(\sigma,K)} \subset c\Sigma_{L(\sigma,K)}$. Using the contractibility of $c\Sigma_{L(\sigma,K)}$, g may easily be extended to the remainder of $\tilde{\sigma}^* \cap \gamma$.

It remains to observe that this extension may be chosen for each γ, so that the definition is consistent, i.e. for $\tilde{\sigma} \subset \delta < \gamma, \gamma'$, the two extensions coincide on $\tilde{\sigma}^* \cap \delta$. Roughly, we see this by noting that in the extension of g above, $g|\tilde{\sigma}^* \cap \delta$ may be chosen to map to a cell of $\mathcal{G}(\delta)$ which is common to all the cells $e_{L(\sigma,K)}$ which arise from the various choices of $(n+k)$-simplices γ. We leave further details to the reader.

We have thus defined g on all of $M_\gamma \cap M_0$ for each γ and therefore on all of M_0 where M_0 as usual is $\bigcup_{\sigma \not\subset \partial M} \sigma^*$. We must extend this map to a PL bundle map $\hat{g}: TM_0 \to \gamma_{n,k}(W)$. This is done by constructing on M a cellular decomposition $\{e_\sigma\}$ of M, where e_σ is a regular neighborhood (in M_0) of the barycenter b_σ of a given simplex σ of M. We omit details, merely pointing out that this is perfectly analogous to the decomposition $\{\bar{e}_L\}$ of $\mathcal{G}_{n,k}$ or $\{\tilde{e}_L^\alpha\}$ of $\mathcal{G}_{n,k}(W)$. So, in particular, under the Gauss map g as defined $g(e_\sigma)$ will lie in $e_{L'}^{\delta'}$ where $\delta' < \delta$ and $L(\sigma, M_\delta) = \iota_{\delta',\delta} L'$ where $L(\sigma, M_\delta)$ denotes the formal link of σ under the immersion $\phi_\delta \circ (f|M_\delta)$. Thus, $G \circ g|e_\sigma$ will be an embedding of e_σ in $V_{L'}^{\delta'}$ and hence there is a bundle map covering this embedding $TM|e_\sigma = Te_\sigma \to TV_{L'}^{\delta'}$. But, locally, $\gamma_{n,k}(W)|\tilde{e}_L^{\delta'}$ is $G^*TV_{L'}^{\delta'}$, and so this bundle map on Te_σ may be construed as a map $\hat{g}$ to $\gamma_{n,k}(W)$ covering g. It remains to note that if $x \in e_\sigma \cap e_\tau$, the two definitions co-incide. Again, details are left to the reader.

In the spirit of §2, we consider <u>geometric</u> subcomplexes of $\mathcal{G}_{n,k}(W)$.

<u>6.4 Definition</u>. A subcomplex $B \subset \mathcal{G}_{n,k}(W)$ is said to be geometic iff given $e_L \subset B$ and $V_{L'} \cap \delta = V_L \cap \delta \neq \emptyset$ for some maximal simplex δ of W then $e_{L'} \subset B$.

As usual, we shall now assume that M^n has no closed components.

<u>6.5 Theorem</u>. Let B be a geometric subcomplex of $\mathcal{G}_{n,k}(W)$ (W with a fixed triangulation).

Assume there is a bundle map

$$\begin{array}{ccc} TM & \xrightarrow{\hat{f}} & Y_{n,k}(W)|B \\ \downarrow & & \downarrow \\ M & \xrightarrow{f} & B \end{array}$$

Then there exists an immersion $F\colon M \to B$ with $F \simeq G \circ f$, and the Gauss map g_F having image in B.

Proof: Consider $\tilde{B} = \bigcup_{e_L \subset B} \tilde{e}_L$. As we have previously remarked, in the case of subcomplexes of $\mathcal{G}_{n,k}$, $B \subset \tilde{B}$ is a homotopy equivalence, and we may thus regard f as a map into $\tilde{B}$.

By the usual general position arguments, we may assume that f is simultaneously transverse to all "strata" of $\tilde{B}$ having the form $\tilde{e} = \tilde{e}_{L_1} \cap \ldots \cap \tilde{e}_{L_r}$ for a finite collection of formal links $\mathcal{L} = \{L_1, \ldots, L_r\}$ (ordered by the face relation).

Let $M_L = f^{-1}(\tilde{e}_L)$; thus M_L is a codimension-0 submanifold of M^n. The idea of the proof is to obtain a codimension-0 immersion $M_L \to \bar{V}_L$, with $M_L \cap M_K \to \bar{V}_L \cap \bar{V}_K$. Let $\mathcal{L} = \{L_1, \ldots, L_r\}$ be a finite set of formal links ordered by the face relation. Then $G(\tilde{e}_{\mathcal{L}})$ is a compact subset of $\operatorname{int}(\bigcap_{L \in \mathcal{L}} \bar{V}_L)$, which is a PL-homeomorph of R^n.

Let $M_{\mathcal{L}}$ denote $\bigcap_{L \in \mathcal{L}} M_L$ and $V_{\mathcal{L}}$ be a regular neighborhood of $G(\tilde{e}_{\mathcal{L}})$ in $\mathrm{int}(\bigcap_{L \in \mathcal{L}} \bar{V}_L)$.

Now note that for $\tilde{e}_{\mathcal{L}} \neq \emptyset$, $\mathcal{L}$ must have size $n+1-j$, $0 \leq j \leq n$. We work inductively, starting with $j = 0$; i.e., the induction is on the increasing dimension d of the manifold $M_{\mathcal{L}}$ (Note that $\dim M_{\mathcal{L}} = j$, j as above). Our inductive assumption is that $\tilde{F} = (G|B) \circ f$ has been deformed to F' so that $F'|M_{\mathcal{L}}$ is an immersion into $V_{\mathcal{L}}$ for all $\mathcal{L}$ of length $\geq n+1-d$, i.e. on all $M_{\mathcal{L}}$ of $\dim \leq d$. We assume that this deformation keeps $M_{\mathcal{L}}$ in $V_{\mathcal{L}}$ throughout ($\mathcal{L}$ of arbitrary length). Moreover, we assume that on each $M_{\mathcal{L}}$ we have a PL bundle map $TM|M_{\mathcal{L}} \xrightarrow{\hat{F}'} TV_{\mathcal{L}}$ covering $F'|M_{\mathcal{L}} \to V_{\mathcal{L}}$, the deformation of F to F' having been covered by a deformation of the original bundle map which covered F, viz. $TM|M_{\mathcal{L}} \xrightarrow{\hat{f}} \gamma_{n,k}(W)|\tilde{e} \to TV_{\mathcal{L}}$. We assume that this family of bundle maps is consistent under restriction from $M_{\mathcal{L}}$ to $M_{\mathcal{J}}$, $\mathcal{L} \subset \mathcal{J}$. Moreover this family of deformations of bundle maps is also taken to respect restrictions. Finally for $\dim M_{\mathcal{L}} \leq d$, we shall assume that the sub-bundle inclusion $TM_{\mathcal{L}} \subset F_{\mathcal{L}}^{*}\, TV_{\mathcal{L}}$ coming from the immersion $F_{\mathcal{L}}'$ coincides with the sub-bundle inclusion $TM_{\mathcal{L}} \subset TM|M_{\mathcal{L}} \xrightarrow{\hat{F}'} TV_{\mathcal{L}}$.

These inductive assumptions in mind, we consider a typical $M_{\mathcal{L}}$ of dimension $d + 1$. If we take $d + 1 < n$, we see that, in view of the relative version of the Hirsch-Poenaru theorem [H-P] that $F_{\mathcal{L}}'$ may be deformed to an immersion $F_{\mathcal{L}}'' : M_{\mathcal{L}} \to V_{\mathcal{L}}$. This deformation will be constant on each $M_{\mathcal{J}}$, $M_{\mathcal{J}} \subset \partial M_{\mathcal{L}}$ and, moreover, will be covered by a deformation of bundle maps from $\hat{f}'$ to $\hat{f}''$ where the latter has the property that $\hat{f}''|TM_{\mathcal{L}}$ is merely the bundle inclusion arising from the immersion $M_{\mathcal{L}} \xrightarrow{F_{\mathcal{L}}''} V_{\mathcal{L}}$. It is understood that this deformation is the trivial one over each $M_{\mathcal{J}} \subset \partial M_{\mathcal{L}}$. Finally, since each deformation of $F_{\mathcal{L}}'$ to $F_{\mathcal{L}}''$ is a deformation of the restriction of the global map

$F: M \to W$, and since the various deformations are constant (and hence agree) on all $M_{\mathcal{J}}$, $\mathcal{J}$ of length $> n+1-d$, it follows that there is a global deformation of F' to $f'': M \to W$, with the property that the deformation keeps $M_{\mathcal{K}}$ in $V_{\mathcal{K}}$ ($\mathcal{K}$ arbitrary), where $F''|M_{\mathcal{J}} = F''_{\mathcal{J}}$, (and similarly for the respective deformations). Likewise the deformation of bundle maps from $\hat{f}'_{\mathcal{J}}$ to $\hat{f}''_{\mathcal{J}}$ extends, on each M, $\dim M > d+1$, to a deformation from $\hat{f}'_{\mathcal{K}}$ to $\hat{f}''_{\mathcal{K}}: TM|M_{\mathcal{K}} \to TV_{\mathcal{K}}$, which, of course, covers the deformation of $F'_{\mathcal{K}}$ to $F''_{\mathcal{K}}$. This, at the end of this inductive step, we may relabel F'' as F' and see that we have attained our original inductive hypothesis, with replaced by $d+1$.

The final step, then, consists of considering the case where $= n$, i.e., our assumption is now that F' is an immersion on each $M_{\mathcal{J}}$, length $\mathcal{J} \geq 2$. So we consider the problem of deforming $F_L = F|M_L$ for each L with $e_L \subset B$. Since $\dim M_L = \dim V_L = n$ we may no longer apply the Hirsch-Poenaru theorem automatically. Rather, we consider $M^\circ_L = M_L$ with a disc removed from the interior of each component. The previous inductive step argument now assures us that, keeping ∂M_L fixed $F'_L|M^\circ_L$ may be deformed to an immersion F''_L into V_L. It is easily seen that this may be done so that $F''_0 = \bigcup_L F''_L$ is a global immersion on all of $M' = \bigcup_L M^\circ_L$. But, as in §2, note that we may find a copy M_1, of M in M', so that $M = M_1 \cup$ (collar). (Here, we use once more the hypothesis that M has no closed components.) Re-parameterizing M_1, as M and $F''_0|M_1$ as F, we obtain the immersion $F: M \to W$. It is obvious that F is homotopic to $G \circ f$.

Now, we assume, perhaps after a slight further deformation, that F is in general position with respect to the given triangulation of W. This may be done, retaining the condition that for each L with $e_L \subset B$, we have a codimension-0 submanifold M_L of M (that is,

$M_1 \cap M_L$ in the old notation) with $F: M_L \to V_L \subset W$ a codimension 0 immersion. Suppose, then, that we triangulate M so that each $M_\sigma = F^{-1}\alpha$, α a simplex of W, is a subcomplex. Let σ be a simplex of M, $\sigma \not\subset \partial M$ in this triangulation, with $b_\sigma \in M_L$, and let β be the smallest simplex of W such that $\sigma \subset M_\beta$. We must show $g_F(\sigma^*) \subset \beta$.

We sketch this fact as follows: Let $\delta > \beta$ be a maximal dimension simplex and K the formal link, $e_K \subset \mathcal{G}_{n,k}(\delta)$ (constructed in the definition of the Gauss map g) so that $\sigma^* \cap \delta$ has image in e_K. Since a regular neighborhood of b_σ goes to $\bar{V}_L$ under F, it follows that $e_L \subset \mathcal{G}_{n,k}(\delta)$. Thus, on the model of the analogous argument in §2, either $\bar{V}_L \cap \delta = \bar{V}_K \cap \delta$ or $\bar{V}_{L'} \cap \delta = \bar{V}_K \cap \delta$ for some face $L' < L$, $e_{L'} \subset \mathcal{G}_{n,k}(\delta)$. (Perhaps another way of saying the same thing is this: If we take $\mathcal{G}_{n,k}(\delta)$ to be identical with the standard $\mathcal{G}_{n,k}$ and use V_L, V_K etc. in their old sense (as in §2)) then V_K is seen to be identical with $V_{L'}$ for some $L' < L$.). Thus, since $e_{L'} \subset \beta$, $e_K \subset \beta$ by the geometricity of β. Hence $g_F(\sigma^* \cap \delta) \subset \beta$. Thus $\operatorname{im} g_F \subset \beta$, which was to be proved.

6.1 Addendum: Equivariant immersions into manifolds

As in the proceeding section, we assume that W^{n+k} is a PL manifold with a fixed triangulation, but with the added hypothesis that the finite group π acts on W as a group of simplicial automorphisms. We shall assume as in §5 that the action of π on W is <u>locally smooth</u> (see, e.g., [Br]). $\mathcal{G}_{n,k}(W)$ therewith becomes a π-space as follows:

Let $u \in \pi$, α^{n+k} a simplex of W with $u(\alpha) = \beta$. The simplicial homeomorphism $u: \alpha \to \beta$ clearly determines an element $\psi_{\alpha,\beta}(u) \in O(n+k)$ with $\psi_{\alpha,\beta}(u)(\Delta^{n+k}) = \Delta^{n+k}$, characterized by

$\psi_{\alpha,\beta}(u)(x) = \phi_\beta \circ u \circ \phi_\alpha^{-1}(x)$, $x \in \Delta^{n+k}$. Here, ϕ_α (resp. ϕ_β) is the homeomorphism $\alpha \to \Delta^{n+k}$ (resp. $\beta \to \Delta^{n+k}$) of the previous section.

<u>6.6 Definition</u>. Let $e_L \subset \mathcal{G}_{n,k}(\alpha)$, where $L = (U_L, \Sigma_L)$. Then, for $u \in \pi$, $u(e_L) = e_{u(L)}$ where $u(L)$ is the formal link given by $u(L) = (\psi_{\alpha,\beta}(u)(U_L),\ \psi_{\alpha,\beta}(u)(\Sigma_L))$.

It should be clear that, since we think of e_L, $e_{u(L)}$ as images, respectively of $c\Sigma_L$, $c\Sigma_{u(L)}$, the map $\psi_{\alpha,\beta}(u)$ induces a homeomorphism $u\colon e_L \to e_{u(L)}$.

We claim, leaving the elementary task of verification to the reader, that the map thus defined, viz, $u\colon \mathcal{G}_{n,k}(\alpha) \to \mathcal{G}_{n,k}(\beta)$ is globally consistent i.e.

<u>6.7 Proposition</u>. If $e_L \subset \mathcal{G}_{n,k}(\alpha) \cap \mathcal{G}_{n,k}(\alpha')$, with $u(\alpha) = \beta$, $u(\alpha') = \beta'$, then $e_{u(L)} \subset \mathcal{G}_{n,k}(\beta) \cap \mathcal{G}_{n,k}(\beta')$, and moreover the map $u\colon e_L \to e_{u(L)}$ is determined the same way by $\psi_{\alpha,\beta}(u)$ as by $\psi_{\alpha',\beta'}(u)$.

Moreover

<u>6.8 Proposition</u>. If $e_L \subset \mathcal{G}_{n,k}(A)$ and $K < L$ then $u(K) < u(L)$ and the diagram

$$\begin{array}{ccc} e_K & \xrightarrow{u} & e_{u(K)} \\ \cap & & \cap \\ e_L & \xrightarrow{u} & e_{u(L)} \end{array}$$

commutes.

Again, the proof is purely routine. Thus since $\mathcal{G}_{n,k}(W) = \bigcup_{\alpha^{n+k}} \mathcal{G}_{n,k}(\alpha)$ we see that a global homeomorphism $u\colon \mathcal{G}_{n,k}(W) \to \mathcal{G}_{n,k}(W)$ arises. Moreover, when $\mathcal{G}_{n,k}(W)$ is realized as a semi-simplicial complex (via the first barycentric subdivision of the cell structure $\{e_L\}$), u is a semisemplicial map.

It is straightforward, as well, to check that for $u, w \in \pi$ $u \circ w = uw$ as self-homeomorphisms of $\mathcal{G}_{n,k}(W)$. Thus $\mathcal{G}_{n,k}(W)$ is seen to be a π-space.

We suppose now that M^n is a PL manifold admitting a locally smooth PL action, and that $f: M \to W$ is a π-equivariant immersion. Suppose further that f is transverse to the given triangulation of W at all points in int M.

This last supposition is by no means vacuous. Not every equivariant immersion is regularly homotopic to one with this transversality property. For instance, let $\pi = Z_2 \oplus Z_2$, and let W be the disc, triangulated as below:

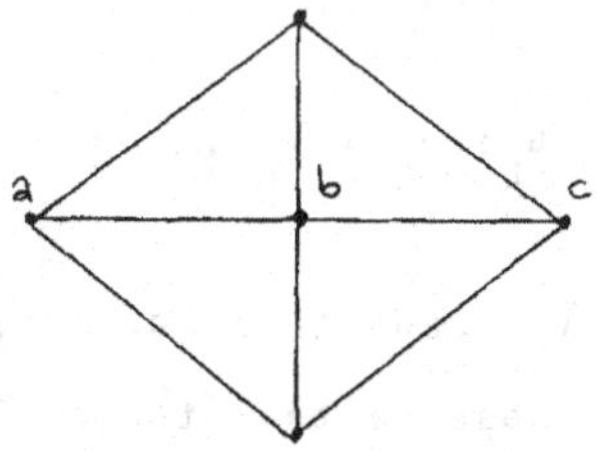

The $Z_2 \oplus Z_2$ action is given by having the generator of one copy of Z_2 flip the figure about the horizontal axis, whilst the generator of the other is the flip about the vertical axis. Let M be two 1-simplices with a common vertex

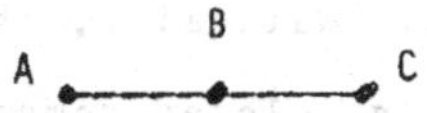

with $Z_2 \oplus Z_2$ action where the first generator acts trivially while the second interchanges A and C, leaving B fixed. The inclusion determined by $A \mapsto a$, $B \mapsto b$, $C \mapsto c$ is clearly equivariant and, equally clearly, is not deformable to an equivariant immersion in general position with respect to the triangulation.

With the hypothesis that $f: M^n \to W^n$ is, in fact, transverse to the triangulation, we examine the Gauss map $g(f): M^n \to \mathcal{G}_{n,k}(W)$ $(M_o^n \to \mathcal{G}_{n,k}(W)$ if M^n has boundary, and conclude, by inspection, that $g(f)$ is a π-equivariant map.

At the same time, we note that the canonical bundle $\gamma_{n,k}(W)$ is, in fact, a π-bundle. This is seen by noting that, locally, as seen previously, the bundle $\gamma_{n,k}(W)$ is induced by the tautological map G_W from the various tangent bundles $T\bar{V}_L^{\alpha}$. Thus, if $e_L \subset \mathcal{G}_{j,k}(\alpha^{j+k})$ but not in $\mathcal{G}_{i,k}(\beta^{i+k})$ for any $\beta < \alpha$, we observe that we may form $\bigcup_{u \in \pi} \bar{V}_{u(L)}^{u(\alpha)} = V_L^{\pi}$ (regarded as a union of disjoint manifolds.) Now let $\tilde{e}_L^{\pi}$ denote the union $\bigcup_{u \in \pi} u(e_L^{\alpha}) \subset \mathcal{G}_{n,k}(W)$. Note that $\tilde{e}_L^{\pi}$ is a π-invariant subspace of $\mathcal{G}_{n,k}(W)$. It is clear that G_W is equivariant, and moreover $G_W|\tilde{e}_L^{\pi} \to W$ factors naturally through a π-equivariant map

$$G_L^{\pi}: \tilde{e}_L^{\pi} \to \bar{V}_L^{\pi},$$

Thus, $\gamma_{n,k}(W)|\tilde{e}_L^{\pi} = (G_L^{\pi})^* \, T\bar{V}_L^{\pi}$, and thus acquires a π-action as a PL bundle. It remains only to observe that these "local" π-bundles cohere to determine a global π-action on $\gamma_{n,k}(W)$.

Furthermore, it will follow from definitions that the covering bundle map $\hat{g}: TM^n \to \gamma_{n,k}(W)$ becomes, under these circumstances, a π-map.

We may thus proceed to find an equivariant analogue to the main result 6.5 previously stated. Naturally, the condition that the manifold to be immersed have no closed components will have to be replaced by the analogous condition in equivariant-map theory, viz., the Bierstone condition.

First, we find a condition on subcomplexes of $\mathcal{G}_{n,k}(W)$ extending the notion of geometricity (Def. 6.4).

<u>6.9 Definition</u>. A subcomplex B of $\mathcal{G}_{n,k}(W)$ is said to be

π-geometric iff B is geometric (in the sense of 6.4) and π-invariant under the given action of π on $\mathcal{G}_{n,k}(W)$.

<u>6.10 Theorem</u>. Let $B \subset \mathcal{G}_{n,k}(W)$ be a π-geometric subcomplex. Let M^n be a manifold with locally-smooth PL π-action, satisfying the Bierstone condition (see §5). If there is an equivariant bundle map

$$\begin{array}{ccc} TM & \xrightarrow{\hat{f}} & \gamma_{n,k}(W) \mid B \\ \downarrow & & \downarrow \\ M & \xrightarrow{f} & B \end{array}$$

then there exists an equivariant immersion $F: M \to W$ such that the induced Gauss map g_F has image in B.

Proof: Given any link L, (associated to some simplex α of W), consider its orbit $\pi \cdot L$.

We set $\bar{V}_{(L)} = \coprod_{K \in \pi \cdot L} \bar{V}_K$, which is, of course a manifold admitting a π-action, as well as a natural π-immersion $\iota_{(L)}: \bar{V}_{(L)} \to W$.

As in the proof of 5.2, we replace B by its equivariant neighborhood $\bar{B} = \bigcup_{e_L \subset B} \bar{e}_L$, and regard f as an equivariant map into $\bar{B}$. Consider the stratification of $\bar{B}$ given by $\{\tilde{e}_{\mathcal{J}}\}$, $\mathcal{J} = (L_1), (L_2) \ldots (L_r)$ where ( ) denotes π-orbit, and $L_1 < L_2 \ldots < L_r$; and $e_{L_1} \subset B$. The fact that f may be made transverse to this stratification via an equivariant deformation mimics the analogous step in §5, i.e. Lemma 5.4. Thus, M is equivariantly decomposed into codimension-0 submanifolds $M_{(L)} = f^{-1}\tilde{e}_{(L)}$ and, as in the proof of 5.2, on $M_{(L)}$, $G_W \circ f$ factors uniquely as $M_{(L)} \xrightarrow{f_{(L)}} \bar{V}_{(L)} \to W$ where the map $\bar{V}_{(L)} \to W$ is given by the natural inclusion on each $\bar{V}_K$, $K \in (L)$.

From here on in, the proof mimics that of 5.2, exploiting the immersion lemma 5.5 The reader may check that details are exactly parallel.

7. The Grassmannian for piecewise smooth immersions

7.1 The space $\mathcal{G}^c_{n,k}$

We have heretofore restricted our attention to problems involving piecewise-linear immersions of a PL manifold M^n into R^{n+k} (or into a triangulated manifold W^{n+k}), and have shown how the complex $\mathcal{G}_{n,k}$ (or the $\mathcal{G}_{n,k}$-"bundle" $\mathcal{G}_{n,k}(W)$) and its geometric subcomplexes are related to geometric restrictions on such immersions. In the subsequent sections we shall switch our focus to <u>piecewise smooth</u> immersions of manifolds M^n in R^{n+k} (or, more generally, into Riemannian manifolds W^{n+k}). Although the manifolds M^n will be PL, in the sense that they admit underlying combinatorial structures, piecewise linear properties, per se, will not play an important role. Rather, we are concerned with immersions $M^n \to R^{n+k}$ (or $M^n \to W^{n+k}$) wherein M^n is stratified by smooth manifolds and the immersion is smooth on each stratum. We shall require on such stratifications a certain technical condition, i.e., if X^j is a stratum and S^{n-j-1} its "linking" sphere, then the stratification induced on S^{n-j-1} by that of M^n shall be a simplicial triangulation. (A simplicial complex is, of course, naturally stratified with the simplices themselves as strata.)

There will be a "Grassmannian" and "Gauss map" appropriate to situations of this sort, i.e., a space $\mathcal{G}^c_{n,k}$ such that, given a piecewise-smooth immersion $M^n \to R^{n+k}$ there ensues a natural map $g: M^n \to \mathcal{G}^c_{n,k}$ naturally covered by $TM^n \to \gamma^c_{n,k}$ where $\gamma^c_{n,k}$ is the "canonical" PL n-disc bundle over $\mathcal{G}^c_{n,k}$. Moreover, we shall see that certain subspaces H of $\mathcal{G}^c_{n,k}$ naturally correspond to restrictions of a geometric nature on immersions. The result, as the reader will undoubtedly anticipate, is that, at least for non-closed manifolds M^n, a bundle map $TM^n \to \gamma_{n,k} | H$ guarantees the existence of a piecewise-smooth immersion $M^n \to R^{n+k}$ whose Gauss map has its image in H. In fact, results of this sort will prove to be a bit stronger than the corresponding results relating PL immersions to geometric subcomplexes of $\mathcal{G}_{n,k}$.

Our first task, then, is to construct $\mathcal{G}^c_{n,k}$ with its canonical bundle $\gamma^c_{n,k}$, and to define the Gauss map for appropriate immersions. Matters are simplified greatly by the fact that $\mathcal{G}^c_{n,k}$, as a set, coincides precisely with $\mathcal{G}_{n,k}$ as previously defined; $\mathcal{G}^c_{n,k}$ is merely $\mathcal{G}_{n,k}$ with a smaller topology.

We define this topology on the underlying point-set as, essentially, a metric topology. That is, we specify the ε-neighborhoods of a typical point in $\mathcal{G}_{n,k}$ for $\varepsilon > 0$. Since $\mathcal{G}^c_{n,k}$ is, pointwise, identical to $\mathcal{G}_{n,k}$ we see that any such x is the image under the identification map of at least one point $x' \in c\Sigma_L \subset R^{n+k}$, where $L = (U_L, \Sigma_L)$ is a formal link of dimension $(n,k;j)$. We shall say that $y \in \mathcal{G}^c_{n,k}$ is within ε of x iff there are representatives $x' \in c\Sigma_L$, $y' \in c\Sigma_K$ of x and y respectively $(\dim L = \dim K = j)$ such that:

i) U_L is within ε of U_K in the standard metric on the standard Grassmannian $G_{j+k,n-j}$

ii) There is a simplicial isomorphism $\phi\colon \Sigma_L \to \Sigma_K$ so that $\phi(v)$ is within ε of v in R^{n+k} for all vertices v of Σ_L

ii) y' is within ε of x' in R^{n+k}.

As usual, $0 \subset \mathcal{G}^c_{n,k}$ is open iff for every $x \in 0$ there is some ε-neighborhood of x, $\mathcal{U}(x)$, with $\mathcal{U}(x) \subset 0$.

Another way of characterizing the topology of $\mathcal{G}^c_{n,k}$ is as follows: the first barycentric subdivision of $\mathcal{G}_{n,k}$ is a simplicial complex, that is, the geometric realization of a simplicial set (i.e., a simplicial space with the discrete topology). Note that each simplex of this particular simplicial complex has a natural linear ordering on its vertices.

If we now retopologize this simplicial space so that the set of j-dimensional simplices has a smaller topology we shall have, in passing to the geometric realization, retopologized $\mathcal{G}_{n,k}$ as well.

Consider, therefore, a typical j-simplex σ of the first barycentric subdivision of $\mathcal{G}_{n,k}$. There is a unique formal link (of dimension r) which we denote by $L(\sigma)$ such that $\text{int}\,\sigma \subset \text{int}\,e_L$. We therefore define an ε-neighborhood of σ in the set of j-simplices by the following: Say that σ' is within ε of σ iff

i) $\dim L(\sigma') = \dim L(\sigma) = r$

ii) $U_{L(\sigma')}$ is within ε of $U_{L(\sigma)}$ in the standard metric on the standard Grassmannian $G_{k+r,n-r}$

iii) There is a simplicial isomorphism $\phi: \Sigma_{L(\sigma)} \to \Sigma_{L(\sigma')}$ such that $\phi(v)$ is within ε of v in R^{n+k} for each vertex v of $L(\sigma)$.

Thus we obtain a neighborhood basis for each element of the set of j-simplices, and consequently a topology on this set. It is not hard to show that, with respect to this topology, face operations are continuous maps. Thus we obtain a simplicial space whose geometric realization is $\mathcal{G}^c_{n,k}$.

We leave it to the reader to verify that the two definitions of the topology of $\mathcal{G}^c_{n,k}$ coincide. Clearly, the forgetful map $\mathcal{G}_{n,k} \to \mathcal{G}^c_{n,k}$, which is the identity on the set level, is continuous.

Our next task is to describe the PL n-plane bundle $\gamma^c_{n,k}$ which is to play the role of the canonical bundle over $\mathcal{G}^c_{n,k}$. Pointwise $\gamma^c_{n,k}$ coincides with $\gamma_{n,k}$, i.e., the fiber of $\gamma^c_{n,k}$ over x is to be identified with the fiber of $\gamma_{n,k}$ over x (regarding x as a point of $\mathcal{G}_{n,k}$). The topology of the total space of $\gamma^c_{n,k}$ is easily described. A point of $\gamma_{n,k}$ lying over x may be specified (according to one of our characterizations of $\gamma_{n,k}$) as the image of a point in the tangent bundle to V_L at $x_0 \in b_L \subset V_L$ where x_0 is a pre-image of x. This means, in effect, that this point in $\gamma_{n,k}$

may be described as the image of a pair (x_o, y_o) where $y_o \in V_L$ is close to x_o. We therefore may characterize the ε-neighborhood of this point (in the topology for $Y^c_{n,k}$) as the set of all points of $Y_{n,k}$ which may be described by pairs x'_o, y'_o where $x'_o \in b_{L'}$ is in the pre-image of x', (x' within ε of x in $\mathcal{G}^c_{n,k}$) and where y'_o is within ε of y_o in R^{n+k}.

With respect to this topology, $Y^c_{n,k}$ is clearly a topological n-disc bundle over $\mathcal{G}^c_{n,k}$. A slight additional argument must be made in order to verify that this bundle admits a natural $PL(n)$ structure. We may see this by first decomposing $\mathcal{G}^c_{n,k}$ into certain closed subsets. Let T denote a triangulation of the $(j-1)$ sphere, $0 < j < n$. Let $\mathcal{L}_T$ denote the set of j-dimensional formal links $L = (U_L, \Sigma_L)$ such that Σ_L is abstractly isomorphic to T. Let $P_T = \bigcup_{L \in \mathcal{L}_T} c\Sigma_L$. Here, the topology is understood to be that induced by the natural map $P_T \to N_T$ where $N_T = \bigcup_{L \in \mathcal{L}_T} e_L \subset \mathcal{G}^c_{n,k}$. Now $\mathcal{L}_T$ itself may be topologized by identifying the point $L \in \mathcal{L}_T$ with the cone point of $c\Sigma_L$ in P_T. I.e., if $L, K \in \mathcal{L}_T$ we have K within ε of L provided that U_K is within ε of U_L as points of $G_{j+k,n-j}$ and $\phi(v)$ is within ε of v for some simplicial isomorphism $\phi\colon \Sigma_L \to \Sigma_K$ and all vertices v of Σ_L. In particular the following is a neighborhood of $L \in \mathcal{L}_T$: Pick a small neighborhood $\mathcal{U}$ of U_L in $G_{j+k,n-j}$. Pick a map $s\colon \mathcal{U} \to O(n+k)$ so that $s(U) \cdot U_L = U$ for $U \in \mathcal{U}$ where $\cdot$ denotes the natural action of $O(n+k)$ on $G_{j+k,n-j}$. Denote the vertices of Σ_L by $v_1, v_2, \ldots, v_q$. For each v_i pick a neighborhood $\mathcal{V}_i$ in the $(j+k-1)$-sphere S_{U_L}. We claim that if $\mathcal{U}$ and the $\mathcal{V}_i$ are chosen to be small enough then $\mathcal{U} \times \mathcal{V}_1 \times \mathcal{V}_2 \times \ldots \times \mathcal{V}_q$ is homeomorphic to a neighborhood of L in ${}_T$. That is, given $w = (U, y_1 \ldots y_q) \in \mathcal{U} \times \mathcal{V}_1 \times \ldots \times \mathcal{V}_q$ define a link L_w by setting $U_{L_w} = U$ and letting Σ_{L_w} be the image of Σ_L under the "geodesic" extension to Σ_L of the assignment $v_i \to s(U) \cdot y_i$.

Clearly, this correspondence is a homeomorphism of $\mathcal{U}\times\mathcal{V}_1\times \times\ldots\times\mathcal{V}_\ell=\mathcal{W}$ onto an open neighborhood of L in $\mathcal{L}_T$. But now observe that $\mathcal{W}$ is a smooth manifold (of a dimension depending on j and on the number of vertices of T). Thus, $\mathcal{L}_T$ is at least a topological manifold and, in fact, a smooth manifold since it is easy to see that these local charts fit together smoothly. For our purposes, it suffices to observe that $\mathcal{L}_T$ is thus a PL manifold. It follows that the space P_T which we have defined above, is, in geometric terms a PL manifold, in fact a PL j-disc bundle over $\mathcal{L}_T$. (The PL structure comes about since P_T is really a bundle over $\mathcal{L}_T$ with structure group given by the simplicial automorphisms of T.)

Let γ_T denote the pullback of $\gamma^c_{n,k}$ to P_T under the map $P_T \to N_T \subset \mathcal{G}^c_{n,k}$. γ_T may easily be seen to be identifiable with a Whitney sum $\zeta\oplus\eta$ where ζ is the PL bundle of "tangents along the fiber" of the j-disc bundle P_T over $\mathcal{L}_T$ and where η is the pullback of an $(n-j)$-vector bundle η_o over $\mathcal{L}_T$, viz; η_o is the bundle whose fiber over $L\in\mathcal{L}_T$ is the vector space $U_L \subset R^{n+k}$.

Not only does γ_T admit a PL structure, but the same is also true of the bundle from which γ_T was originally induced, viz, $\gamma^c_{n,k}|N_T$. The point is that if we view $\gamma_T \to \gamma^c_{n,k}|N_T$ as a quotient map, the identifications which produce it glue fibers together by PL isomorphisms.

Thus, we have specified particular PL structures for the various restrictions $\gamma^c_{n,k}|N_T$. Now suppose $N_{T_1} \cap N_{T_2} \neq \emptyset$. This only occurs when T_1 is the link of a simplex T_2 (or vice-versa). We claim that the two PL structures on $\gamma^c_{n,k}|N_1 \cap N_2$ coincide. We leave detailed verification of this point to the reader as a straightforward exercise. Since $\{N_T\}$ covers $\mathcal{G}^c_{n,k}$ we have in fact specified a PL structure for $\gamma^c_{n,k}$.

We now wish to analyze further some aspects of the geometric structure of the space $\mathcal{G}^c_{n,k}$. We have already introduced the closed subspace N_T and we have noted that union in N_T = image $\bigcup_{L\in\mathcal{L}_T} c\Sigma_L$ of all cone points may be identified with $\mathcal{L}_T$. Note also that N_T - image $\bigcup_{L\in\mathcal{L}_T}\Sigma_L$ is a homeomorph of int $P_T = \bigcup_{L\in\mathcal{L}_T} \mathring{c}\,\Sigma_L$ where $\mathring{c}$ denotes open cone. Let c' denote the cone of "radius" $\frac{1}{2}$ inside the cone c of "radius" 1, and let O_T N_T be the homeomorphic image of $\bigcup_{L\in\mathcal{L}_T} c'\Sigma_L$. (Thus O_T is abstractly homeomorphic to P_T as a space and equivalent as a disc bundle over $\mathcal{L}_T$.)

Recall that there is a standard vector bundle η_T over $\mathcal{L}_T$ whose fiber over $L \in \mathcal{L}_T$, $L = (U_L, \Sigma_L)$, is the vector subspace U_L of R^{n+k} ($\dim \eta_T = n-j$ when $\dim L = j$ for all $L \in \mathcal{L}_T$). Obviously, then, there is a natural map θ_T from $\mathcal{L}_T$ to the standard Grassmannian $G_{n-j,j+k}$ classifying η_T and explicitly given by $\theta_T: L \to U_L \in G_{n-j,j+k}$.

<u>7.1 Lemma</u>. θ_T is a fibration (in the sense of Serre).

Proof: Let X be an arbitrary finite complex and let $f: X \to \mathcal{L}_T$ be an arbitrary map, $g = \theta_T \circ f$ and $g: X\times I \to G_{n-j,k+j}$ a homotopy with $G_0 = g$. We must exhibit a homotopy $F: X\times I \to \mathcal{L}_T$ with $F_0 = f$ and $\theta_T \circ F = G$.

Consider the standard fibering $V_{n-j,j+k} \xrightarrow{\pi} G_{n-j,j+k}$ of the Stieffel manifold over the Grassmannian. Choose a covering $\{A_i\}$ of $G_{n-j,j+k}$ such that there are local sections $s_i: A_i \to V_{n-j,j+k}$ of π. Pick a subdivision of X and a partition $0 = t_0 < t_1 < \ldots < t_q = 1$ of I so that $\theta_T(\sigma\times[t_r,t_{r+1}]) \subset A_i$ for at least one i where σ is an arbitrary simplex of the subdivided X and $r < q$. Over each such $\sigma\times[t_r,t_{r+1}]$ we have $\alpha = s_i \circ G$, i.e. $\pi\circ\alpha = G$. Assume, inductively that F has been defined on:

$$X\times\{0\} \cup X^{(p)}\times I \cup X^{(p+1)}\times[0,t_r].$$

(Here, $X^{(p)}$, $X^{(p+1)}$ denote the p and $p+1$ skeleta respectively (in the subdivision) and $r < q$.) Consider $\sigma\times[t_r,t_{r+1}]$ and re-parameterize it as $\sigma'\times I$ where $\sigma'\times\{0\}$ is identified with $\sigma\times\{t_r\}\subset\sigma\times[t_r,t_{r+1}]$. Define F on $\sigma'\times I$ as $F(x,t) = (U_{(x,t)},\sum_{(x,t)})$ for each $x\in\sigma'$, $t\in I$ by the following procedure: First of all, $U_{(x,t)} = (G(\overline{x,t}))$ where $(\overline{x,t})$ corresponds to (x,t) in the original parameterization of $\sigma\times[t_r,t_{r+1}]$. To define $\sum_{(x,t)}$ note that $\alpha(\overline{x,t})$, $\alpha(\overline{x,0})$ give ordered orthonormal bases for $U_{(x,t)}$, $U_{(x,0)}$ respectively, and thus a continuous (in x and t) family of isometries $\Phi_{(x,t)}: U_{(x,t)} \to U_{(x,0)}$. Thus we may define $\sum_{(x,t)}$ as $\Phi^{-1}_{(x,t)}(\sum_{(x,0)})$ (with the obvious simplicial structure). Thus, F has been defined on $\sigma'\times I = \sigma\times[t_r,t_{r+1}]$ with $\theta_T\circ F = G$. Hence, since σ was arbitrary, we have extended F to $X\times\{0\}\cup X^{(p)}\times I\cup X^{(p+1)}\times[0,t_{r+1}]$ and the most routine of inductive arguments show that we may extend F to $X\times\{0\}\cup X^{(p+1)}\times I$ and therefore, finally, to all of $X\times I$. The proof of 7.1 is thus complete.

(In fact, one might also observe that $\mathcal{L}_T$ is a locally-trivial fiber bundle over $G_{n-j,j+k}$ with projection map θ_T whose fiber, determined by T, is as follows: Order the vertices of T in some fashion as $v_1,v_2,\ldots,v_s$ and consider the space of maps $\{v_1,\ldots,v_s\}\to S^{j+k-1}$ which induce piecewise-geodesic embeddings of T, the topology being induced from the s-fold cartesian product of S^{j+k-1} with itself. Identify two points in this space if they differ by a permutation of the v's which extends to an automorphism of T. The identification space is, in fact, the fiber of $\theta_T:\mathcal{L}_T\to G_{n-j,j+k}$.)

We note in passing that if 0 denotes the unique "triangulation" of $S^{-1} = \emptyset$, then $\mathcal{L}_0$ = image of 0-skeleton of $\mathcal{G}_{n,k}$ is, in fact, a copy of the standard Grassmannian $G_{n,k}$ embedded in $\mathcal{G}^c_{n,k}$.

$\gamma_{n,k}|\mathcal{L}_0$ is the standard n-vector bundle over $G_{n,k}$ and θ_0 is the identity.

We shall show, in the next two sections, that there are certain geometric situations giving rise to a "Gauss map" into $\mathcal{G}^c_{n,k}$. These involve "piecewise differentiable" immersions of manifolds M^n into R^{n+k}.

7.2 LS-Stratified Manifolds

Consider a closed PL manifold M^n. A <u>strict stratification</u> of M^n shall denote, specifically, a decomposition of M^n into closed, connected subspaces X_i called <u>strata</u> such that

1) Each X_i is a (locally flat) compact PL submanifold of M^n

2) ∂X_i is the union of strata of lower dimension

3) If X_1, X_2 are distinct strata, then int X_1 is disjoint from int X_2. Moreover, $X_1 \cap X_2 \neq \emptyset$ only if $X_1 \subset \partial X_2$ or $X_2 \subset \partial X_1$ or $\partial X_1 \cap \partial X_2$ is the union of lower dimensional strata.

We shall say that X_j is incident to X_i (notation: $X_j < X_i$) iff $X_j \subset X_i$. The symbol $\leq$ means incident or equal to.

The extension of the notion of strict stratification to manifolds M^n with non-void boundary or to open manifolds is immediate. In the case of compact manifolds with boundary one merely insists that the strata of M^n meet ∂M^n transversally, resulting in a stratification of ∂M^n where a typical stratum Y is a component of $X \quad \partial M^n$, where X is a stratum of M with codimension of X in M = codimension of Y in ∂M. In the case of an open manifold M^n, one insists that the strata X_i are proper in the sense that $X_i \cap C$ is compact for all compact subspaces C of M.

We shall, for the moment, assume that M^n is a compact, strictly stratified manifold. Let X be a stratum and X_0 any codimension-0 submanifold of X (without boundary) so that $\bar{X}_o$ is disjoint from ∂X. If $X < Y$ and $\dim Y = \dim X+1$ then obviously X_o has a neighborhood in Y in the form of a collar, i.e. a homeomorph of $X_o \times I$. We shall call such a regular neighborhood "good." Now consider a stratum Y, $X < Y$ and $\dim Y - \dim X = q$. We shall call a regular neighborhood R of X_o in Y "good" if it satisfies the following inductively-defined condition viz; $R \subset Y$ is of the form $X_o \times R^q_+$ (where R^q_+ denotes the standard half-space), and $R \cap Z$ is good for any stratum Z with $X < Z < Y$. Finally, we shall call a regular neighborhood R of X_o in M^n "good" iff $R \cap Y$ is good for all Y with $X < Y$.

Good regular neighborhoods clearly exist for any X_o; in particular, it is clear that such a good regular neighborhood R of X_o is strictly stratified where the strata consist of the components of X_o and the components of $R \cap Y$ for all Y with $X < Y$. Furthermore, it is not at all hard to see that, as a strictly-stratified manifold, R has the form $X_o \times \mathcal{D}$, where $\mathcal{D}$ is a stratified disc, $\dim \mathcal{D} = \operatorname{codim} X$ and $\mathcal{D}$ is stratified by dint of taking the cone on a strict stratification of $\dot{\mathcal{D}}$. $\mathcal{D}$ is described in essence by taking a small disc D^j transverse to X at $p \in X_o \cap D^j$ ($j = \operatorname{codim} X$) so that for $X < Y$, $\dim Y - \dim X = q$, we have $D^j \cap Y$ of the form R^q_+. $\dot{D}^j$ is thus stratified strictly with strata $\dot{D}^j \cap Y$ for $X < Y$ and D^j itself is strictly stratified by letting the strata of D^j be the cones on the strata of $\dot{D}^j$, with the cone point deleted, plus the cone point itself as a separate minimal stratum.

Note that this strictly-stratified disc $\mathcal{D}$ depends (in the category of strictly-stratified manifolds) only on X as it sits in the particular strictly-stratified M^n, and not at all on X_o. In particular we have a stratified $(j-1)$-sphere $\dot{\mathcal{D}} = \mathcal{L}(X)$ which shall henceforth be called the <u>link</u> of the stratum X.

The most obvious example of a strictly stratified manifold, of course, is a combinatorially triangulated manifold M^n, where the strata are the simplices. Thus, if σ is a simplex, then $\mathcal{L}(\sigma)$ is the usual link $\ell k(\sigma, M^n)$ which is a triangulated, hence strictly stratified, sphere.

Having defined the notion of <u>strictly stratified manifold</u>, we generalize it slightly and speak of <u>stratified</u> manifolds, meaning, essentially, manifolds which are locally strictly stratified. More precisely, a manifold is said to be stratified iff it is covered by open sets U_i which are strictly stratified so that on $U_i \cap U_j$ the strict stratifications inherited, respectively, from U_i and U_j are identical. In the case of a stratified manifold, we shall say that X is a stratum iff X is connected and $X \cap U_i$ is the union of same-dimensional strata for each U_i in the aforementioned atlas. Thus $\text{int } X = \bigcup_i \text{int } X \quad U_i$ is a manifold and X is a manifold-with-boundary modulo identifications on the boundary.

We illustrate the difference between strictly-stratified and stratified manifolds by means of a simple example.

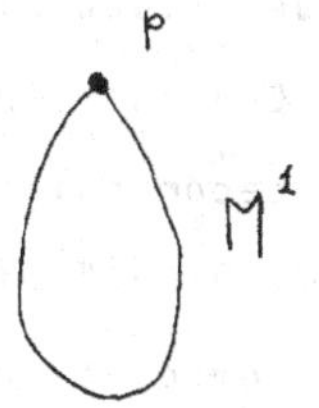

Fig. 7.1

In Figure 7.1 M^1 is a stratified manifold where the strata are the point p (of dimension 0) and the entire circle (of dimension 1). M^1 is stratified because we have a covering by two charts as in Figure 7.2

Fig. 7.1a

Fig. 7.2b

both of which are strictly stratified. However, M^1 itself is not strictly stratified in that the interior of the 1-dimensional stratum which has no boundary (as a manifold), contains the 0-dimensional stratum.

It is clear that even in the case of a manifold which is stratified non-strictly the link $\mathcal{L}(X)$ of a stratum is, up to equivalence, a well-defined strictly-stratified manifold. We may therefore now narrow the class of stratified manifolds under consideration by placing restrictions on the links $\mathcal{L}(X)$ which will be allowed. In particular, we shall call a stratified manifold <u>linkwise simplicial</u> (abbreviated LS) iff $\mathcal{L}(X)$ is equivalent, in the sense of strictly stratified manifolds, to a triangulated sphere (for all strata X.)

The construction about to be described will be of some importance in the subsequent sections. Consider an LS stratified manifold M^n. We shall describe a decomposition of M^n into codimension-0 submanifolds, each of which "thickens" a particular stratum. We shall denote such a decomposition by $\mathcal{M} = \{M(X)\}$ where X ranges over the strata.

First of all, let Σ denote a triangulated sphere of dimension $j-1$. Assume, for convenience sake, that Σ is admissibly embedded in S^{j+k-1}. We shall decompose $c\Sigma$ into codimension-0 submanifolds, namely one N_σ for each simplex σ of Σ, plus a single N_* corresponding to the cone point. This last is merely the smaller cone $c'\Sigma \subset c\Sigma$. For the remaining N's, i.e., the N_σ, assume, inductively that N_σ has been defined for all σ of dimension

$i < j-1$. We must define N_σ for i-dimensional σ, and we do so by first considering the copy Σ' of Σ parameterized as $\Sigma\times\{\frac{1}{2}\}$ in $c\Sigma$ (i.e., $\Sigma' = \partial N_*$). Let σ' be the copy of σ in Σ' for simplices σ of Σ. Let $\bar{\sigma}$ denote the $(j-i-1)$-cell dual to σ' in Σ'; let P denote the $(i+1)$-plane in R^{j+k} determined by σ and the origin and let U be the image of $\bar{\sigma}$ under orthogonal projection onto $P^{\perp}$; finally, let 0 denote the space $\{(x,y) \in P \oplus P = R^{j+k} \mid y \in U\}$. Then N_σ is precisely defined as the closure of the following set:

$$0 \cap c(\text{st } \sigma) - \bigcup_{\dim \tau < i} N_\tau - N_*$$

The following Figure 7.3 illustrates this decomposition when $j = 2$, $k = 0$ and Σ a triangle with vertices v_1, v_2, v_3 and sides s_1, s_2, s_3.

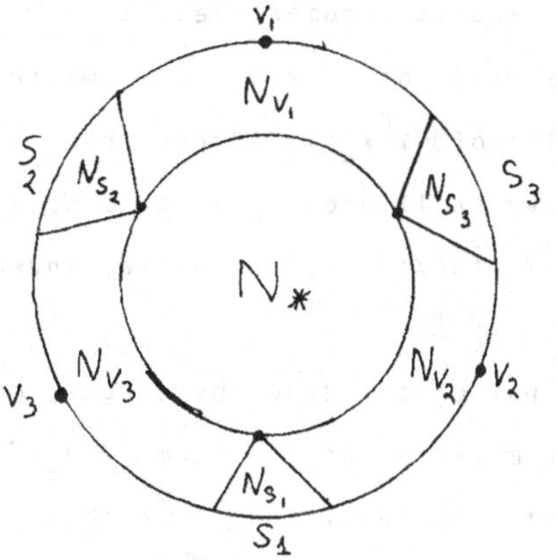

Fig. 7.3

We remark that the assumption that Σ was admissibly triangulated as a subsphere of some standard S^{j+k-1} makes the characterization of N_*, N_σ appear to depend on the embedding. However, it is a fact that, up to isotopy, the simplicial structure of Σ itself determines this decomposition.

Suppose we are given an unstratified connected manifold V^r and a triangulated sphere Σ^{n-r-1}. Let $\text{Aut}(\Sigma)$ denote the group of simplicial automorphisms of Σ and η an $\text{Aut}(\Sigma)$ bundle over V^r.

Associated to η is a bundle with fiber $c\Sigma$, in the obvious way, and we simply denote this as a twisted product $V^r \times_t c\Sigma$, the underlying principal bundle η being understood. $V \times_t c\Sigma$ is a stratified manifold in a natural way. The minimal stratum is $V^r \times_t * = V^r \times * \cong V^r$. The other strata may be described as fiber bundles (i.e., twisted products) over V^r. That is, consider an arbitrary point $v \in V^r$; given a simplex σ of Σ, we then have an action of $\pi_1(V,v)$ on Σ coming from the holonomy $\pi_1(V,v) \to \mathrm{Aut}(\Sigma)$ of the bundle η. Let $O(\sigma)$ be the orbit of the subspace $\mathring{\sigma}$ under this action, and consider $cO(\sigma)-* = X_\sigma \subset c\Sigma$. A typical stratum of $V^r \times_r c\Sigma$ is of the form $V^r \times_t X_\sigma$. Further given $N_* \subset c\Sigma$, we see that we have a neighborhood $V^r \times_t N_*$ of V^r, which we denote $N(V)$. Moreover, given σ in Σ, let $P(\sigma)$ denote the orbit of N_σ under the action of $\pi_1(V,v)$ and it becomes clear that we may define a subtwisted product $V^r \times_t P(\sigma)$ of $V^r \times_t \Sigma$. As a matter of notation, if we take a typical stratum of $V^r \times_t c\Sigma$ (Other than V^r itself) and denote it by X, then we shall use $N(X)$ to denote $V^r \times_t P(\sigma)$ for any σ such that $X = X_\sigma$ ($N(X)$ is, of course, independent of the choice of σ).

We now describe a way of building, by stages, more complicated stratified manifolds from pieces of the form $V^r \times_t c\Sigma$, and, incidentally, obtaining a certain decomposition $\mathcal{M} = \{M(X)\}$ of the finished product.

The 0th stage involves taking various triangulated $(n-1)$-spheres Σ_i and forming the disjoint union $\bigcup_i c\Sigma_i$, regarded as a stratified manifold (with boundary). Call this manifold M_0. We have a decomposition of M_0 into codimension-0 submanifolds $M_0(X)$, one for each stratum of X by noting that any such stratum is of the form $*$ (for one of the $c\Sigma_i$) or $c\sigma-*$ (for σ a simplex of one of the Σ_i). Then $M(X)$ is of the form N_* (in the appropriate $c\Sigma_i$) or N_σ. (Note that $M_0 = 0$ is also a possibility.)

The next stage involves taking 1-manifolds V_i^1 and twisted products $V_i^1 \times_t c\Sigma_i$ for corresponding triangulated $(n-2)$-spheres Σ_i. The disjoint union of these twisted products is a stratified manifold, and, as we have seen, for each stratum X of this union we have a manifold $N(X)$. Now suppose we are given some points in $\partial(\cup V_i)$ (call this set D). Then the restriction of the various twisted products to the points of D gives a stratified $(n-1)$-manifold Δ. For each stratum Y of Δ we have a co-dimension 0 submanifold $N(Y)$, given by the unique component $M(Y)$ of $M(X) \cap \Delta$ such that i) X is a stratum of $\bigcup V_i \times_t c\Sigma_i$ with $X \cap \Delta \supset Y$; ii) $M(Y) \cap Y \neq \emptyset$. By the same token for every stratum Y' of ∂M_0, we have a submanifold $M(Y')$ of ∂M_0. Now let ϕ be a stratification-preserving PL homeomorphism of Δ into ∂M_0 so that $\phi(M(Y)) \subset M(Y')$ whenever $\phi(Y) \subset Y'$. Moreover, if $p \in D$, then we posit that $\phi(M(p)) = M(\phi(p))$. We then may form the union $M_0 \cup_\phi (\bigcup V_i \times_t c\Sigma_i) = M_1$ which is, in an obvious way, a stratified manifold. That is, the strata of M_1 are unions of strata of M_0 and $\bigcup V_i \times_t c\Sigma_i$ such that the union is connected: If such a stratum X is the union of strata $X = \bigcup Y_r$ set $M(X) = \bigcup M(Y_r)$. We thus obtain a decomposition of M_1 into codimension-0 submanifolds.

In general, we proceed inductively and assume that at the rth stage, $r < n$, we have a stratified manifold M_r and a decomposition in the form $\{M(X)\}$, X ranging among the strata of M_r. Suppose we now take a disjoint union $\bigcup V_i^{r+1} \times_t c\Sigma_i$ where the Σ_i are triangulated $(n-r-2)$-spheres. As before, let D be a codimension-0 submanifold of $\bigcup V_i$ and Δ the union of the restrictions of the various twisted products to the components of D. Just as in the building of M_1, we note the decomposition $\{M(Y)\}$ of ∂M_r, Y a stratum of ∂M_r. We consider a stratum-preserving PL homeomorphism ϕ taking Δ into ∂M_r, and form $M_{r+1} = M_r \cup_\phi (\bigcup V_i \times_t c\Sigma_i)$, with its stratification and decomposition. We may continue this process up to the nth stage. (Although we include the possibility that, say, the first p stages, and the last q stages involve adjunction of the null set.) Note that, if we consider the stratification of $M = M_n$ and decomposition $\{M(X)\}$ which has been obtained, and set $X_0 = X \cap M(X)$, it is seen that X_0 is a codimension-0 submanifold of X such that int X_0 is a smaller copy of int X. Moreover, $M(X)$ is a disc-bundle over X_0 with structure group equal to the simplicial automorphisms of the link of X.

We now put forward, without detailed proof, the claim that any LS stratified manifold may be obtained by the procedure described above. (N.B.: we do not exclude the case where an initial sequence of stages, say 0 through r, $r < n$, are trivial, i.e. involve successive adjunctions of the null set. This corresponds to the case where the minimal stratum is of dimension $r+1$.) It follows that any LS-stratified manifold admits a decomposition $\{M(X)\}$. We put forth the further claim that this decomposition is unique, up to ambient isotopy. That is, it is essentially determined by the stratification itself, and not by choices to be made in the course of the iterative procedure. We leave verification of these claims to the reader as a straightforward, albeit somewhat tedious exercise.

By way of fixing ideas, the reader may want to keep in mind, as a motivating example, the special case where the stratification on M^n comes from a combinatorial triangulation, i.e. the strata are the open simplices. In this case, the construction above amounts to a slight elaboration of the procedure for building M^n as a handlebody with one r-handle for each r-simplex. In this case, if X is an r-simplex, then $M(X)$ may be thought of as the actual r-handle with core disc X_0, which is a closed r-simplex (i.e. a "shrinking" of X).

The following simple example may also prove useful as an aid to visualization. Figure 7.4 illustrates a 2-manifold with some distinguished points (0-strata) and arcs (1-strata). The remainder consists of connected 2-manifolds (2-strata).

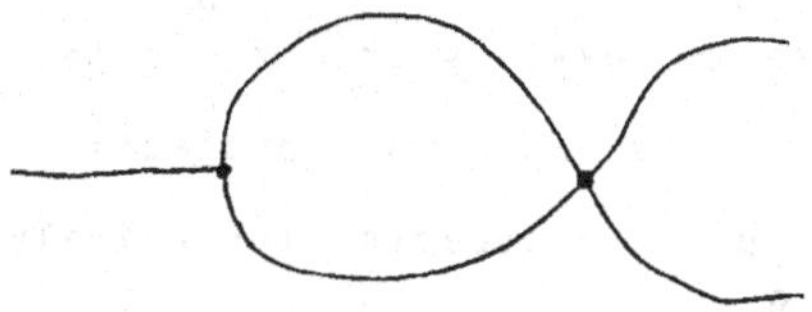

Fig. 7.4

In figure 7.5, the regions bounded by dotted lines are the $M(X)$, and the intersections with corresponding strata the X_o.

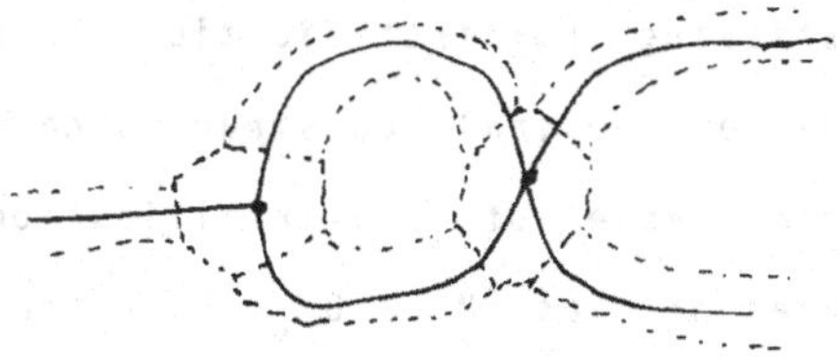

Fig. 7.5

Given the LS-stratified manifold M^n, we are now going to describe a certain map $u: M^n \to M^n$, homotopic to the identity, which preserves closures of strata and has the property that $u: \text{int } X_o \to \text{int } X$ homeomorphically. We proceed inductively; at the rth stage we shall have described u on $\bigcup_{\dim X<r} M(X)$. First of all at the 0th stage, we describe u on each $M(X)$, X a 0-stratum, by the trivial map $M(X) \to X$. Now for X a 1-stratum, u is defined on the closure of $X-X_o$ and obviously extends to a map $X_o \to X$ which restricts to a homeomorphism $\text{int } X_o \to \text{int } X$. This extension defines u on X. Moreover we have a projection $\pi: M(X) \to X_o$, we we define u on $M(X)$ by the composition $u \circ \pi$.

In general, assume that u has been defined on $\bigcup_{\dim Y<r} M(Y)$; we wish to extend u to a typical $M(X)$, where X is a stratum of

dimension r. Again, we note that as it has been defined on the closure of $X-X_o$, u extends, in a rather natural way, to $u: X_o \to X$ inducing a homeomorphism $\operatorname{int} X_o \to \operatorname{int} X$. For a suitable parameterization of $M(X)$ as a fiber bundle over X_o, it is seen that u on $M(X)|Z$ is the same as $u\circ\pi$, where Z is that part of ∂X_o meeting $\bigcup_{\dim Y<r} M(Y)$ and π is the bundle projection map. So we may define u on $M(X)$ as $u\circ\pi$. Proceeding inductively, we finally obtain u on all of M^n.

The usefulness of the map u will become clear in the subsequent section §7.3 where it is critical to the definition of a Gauss map for a certain class of piecewise-differentiable immersions. Prior to beginning that discussion, however, we shall digress briefly to consider the problem of classifying LS-stratifications on a given PL manifold M^n, up to concordance. We shall construct a certain space B_{LS} such that concordance classes of LS-stratifications of M^n correspond to homotopy classes of maps $M^n \to B_{LS}$. Our construction of B_{LS} is largely in the spirit of our previous constructions.

First, for fixed j, consider the set of all admissibly-triangulated $(j-1)$-subspheres Σ of the standard Euclidean sphere S^∞ (i.e. $\Sigma \subset S^{j+k-1} \subset R^{j+k}$ for sufficiently large k).

If Σ is such a sphere and v is a vertex of Σ, we denote by Σ_v the following admissibly triangulated $(j-2)$-sphere: Consider the segment from 0 to v in R^{j+k} and the $(j+k-1)$-plane U orthogonal to this segment and passing through the midpoint. Let S' be a small $(j+k-2)$-sphere in U centered at this midpoint, and set $\Sigma' = S' \cap c\Sigma$. Note that Σ' is a triangulated $(j-2)$-sphere isomorphic to $\ell k(v,\Sigma)$. By the usual translation followed by dilation map S' onto a $(j+k-2)$-sphere S'' of radius 1 centered at the origin, thus obtaining the image Σ_v of Σ'. Clearly Σ_v is an admissibly-triangulated subsphere of $S^{j+k-1} \subset S^\infty$. The procedure may be generalized just as in §2 so that for any r-simplex σ of Σ we obtain the admissibly-triangulated $(j-r-2)$-sphere Σ_σ. We shall write $T < \Sigma$ whenever $T = \Sigma_\sigma$ for some simplex σ of Σ. Note that as as in our construction of $\mathcal{G}_{n,k}$ in §2, there are natural maps $h(\Sigma,\sigma)$: $c\Sigma_\sigma \to \sigma^* \subset \Sigma$ (where σ^* is the dual cell to σ). These maps have the usual consistency properties and thus we may form a cell complex $\hat{B}$ with one j-cell e_Σ = image $c\Sigma$ for each $(j-1)$-sphere Σ, by identifying $c\Sigma_\sigma$ with σ^* under $h(\Sigma,\sigma)$. Note that this complex has a single 0-cell corresponding to the unique "embedding" of $S^{-1} = \emptyset$ in S^∞.

Note that there is a natural forgetful map $\hat{s}: \mathcal{G}_{n,k} \to \hat{B}$ which arises from forgetfully assigning to the formal link (U,Σ) the admissably-triangulated sphere Σ, regarded as a subsphere of S^∞. In fact, this forgetful map is consistent with the natural maps $\mathcal{G}_{n,k} \xrightarrow{\alpha} \mathcal{G}_{n,k+1}$ and $\mathcal{G}_{n,k} \xrightarrow{\beta} \mathcal{G}_{n+1,k}$ so that we obtain a natural forgetful map $\hat{s}$: $\lim_{n,k} \mathcal{G}_{n,k} \to B$.

We now retopologize B (just as $\mathcal{G}_{n,k}$ was retopologized to obtain $\mathcal{G}^c_{n,k}$.) That is, if Σ_1, Σ_2 are admissibly-triangulated $(j-1)$-subspheres of S^∞ (i.e. of S^{j+k} for large k) let them be considered ε-close iff there is a simplicial isomorphism $g: \Sigma_1 \to \Sigma_2$ such that $h(v)$ is within ε of v (with respect to the usual Euclidean metric) for all vertices v of Σ_1. (In passing, note that if T is an equivalence class of triangulations of S^{j-1}, this procedure puts a topology on B_T = the set of all Σ with $\Sigma \in T$.)

In the spirit of the retopologization of $\mathcal{G}_{n,k}$ to obtain $\mathcal{G}^c_{n,k}$, we may now retopologize B as the geometric realization of a simplicial space. I.e., $x,y \in B$ are now to be considered ε-close if $x \in e_{\Sigma_1}$, $y \in e_{\Sigma_2}$ and Σ_1 is ε-close to Σ_2 while $\bar{x}$ is ε-close to $\bar{y}$ in Euclidean space where $\bar{x},\bar{y}$ are pre-images of x any in $c\Sigma_1$ and $c\Sigma_2$ respectively. This retopologization of B is the space which we shall designate B_{LS}. Note that B_T may be identified with a subspace of B_{LS}, namely, the union of the cone points of all $c\Sigma$ such that $\Sigma \in T$. Moreover, the natural map $s: \mathcal{G}_{n,k} \to B$ passes over to a continuous map $s: \mathcal{G}^c_{n,k} \to B_{LS}$. In fact, passing to the limit, we obtain $s: \lim_{n,k} \mathcal{G}^c_{n,k} \to B_{LS}$.

It remains for us to justify the claim that B_{LS} is the universal classifying for LS-stratified structures on PL manifolds. We first determine some additional facts concerning B_T, T a triangulation of the $(j-1)$-sphere. Given T, pick a representative K, i.e. a specific, concrete triangulated sphere representing T. Let E_T be the space of all piecewise-geodesic embeddings of K into S^∞ so that the image is an admissible-triangulated subsphere. Clearly, all such embeddings are determined by their values on the vertices of K, so we may topologize E_T as a subset of $S^\infty \times \ldots \times S^\infty$ where the number of factors is the number of vertices of K.

<u>7.2 Lemma</u>. E_T is weakly contractible.

Proof: Let $\alpha: S^i \to E_T$. We may think of α as parameterized family of admissible embeddings $\phi_x: K \subset S^\infty$, $x \in S^i$. We must show that α is contractible over E_T. Without loss of generality we may assume that $\phi_x: K \to S^N$ for some large fixed N independent of x. Let p denote the number of vertices of K, and order those vertices arbitrarily as $v_1, \ldots, v_p$. Consider $R^{N+p+1} = R^{N+1} \times R^p$ and let n_i be the vector $(0, q_i) \in R^{N+1} \times R^p$ where q_i is the ith standard basis vector of R^p. The assignment $v_i \to n_i$ defines a unique admissible embedding $\phi: K \subset S^{N+p}$. We now define a deformation $\phi_{x,t}$, $0 < t < 1$ such that $\phi_{x,0} = \phi_x$, $\phi_{x,1} = \phi$. It suffices to specify how $\phi_{x,t}$ behaves on the vertices of K. (It will be clear that this specification will define, for each t, an admissible embedding of K). The formula is: $\phi_{x,t}(v_i) = (1-t)\phi_x(v_i) + t \cdot n_i / \| \;\|$ where the symbol $\| \;\|$ in the denominator denotes the usual Euclidean norm of the numerator. 7.2 is thus established.

Now let A_T be the group of simplicial automorphisms of K. Clearly, A_T acts freely on E_T by $a \cdot \phi = \phi \circ a^{-1}$. Thus E_T/A_T is the classifying space B_{A_T} for the finite group A_T. But it is apparent that any element of E_T/A_T is characterized precisely by specifying an admissibly-triangulated $(j-1)$-sphere Σ in S^∞ with $\Sigma \in T$. I.e., there is a natural identification of $B_{A_T} = E_T/A_T$ with $B_T \subset B_{LS}$.

Now, let us consider $D_T = \bigcup_{\Sigma \in T} c\Sigma_j$ with the appropriate topology. This is precisely the universal D^j bundle over B_T with structure group A_T. Note that, as B_{LS} has been constructed, D_T is the pre-image of $\bigcup_{\Sigma \in T} e_\Sigma$. Under the identification map, $D_T - S_T$ goes homeomorphically onto its image, where S_T denotes the boundary sphere bundle of D_T.

Let M_0^n, M_1^n be two LS-stratified structures on the same underlying PL manifold M^n.

<u>7.3 Definition</u>. M_0^n and M_1^n are said to be concordant LS-stratified structures on M iff there is an LS-stratified structure on $M \times I$ so that M_i coincides with that induced on $M \times \{i\}$, $i = 0,1$.

<u>7.4 Lemma</u>. B_{LS} is the universal classifying space for LS-stratified structures on PL manifolds. I.e., for any PL manifold M^n, concordance classes of LS-stratified structures on M^n are in 1-1 correspondence with $[M^n, B_{LS}]$.

The proof is rather straightforward. However, inasmuch as part of the argument will be needed for the proof of the main theorem 7.5, we shall set it out in some detail at this point.

First of all, we shall "stratify" B_{LS} into non-manifold strata X^T, one for each equivalence class of triangulations of a finite-dimensional sphere. Each X^T is to have an open neighborhood looking like an open disc bundle with structure group A_T. As usual, we proceed inductively. At the 0th stage, we take the unique 0-cell, and denote this X_o^o. This is to be part of X^o, where the o in superscript denotes the "triangulation" of $S^{-1} = \emptyset$. At the first stage, we look at the unique triangulation of S^o, which we denote "1" in context. Consider the image in B_{LS} of $\bigcup_\Sigma c\Sigma-*$, Σ ranging over the 0-spheres in S^∞. This image contains X_o^o and we denote it by X_1^o. We let $X_1^1 = B_1$ be the union of all the cone points of the Σ.

Now suppose we have reached the jth stage, i.e. we have stratified $\bigcup_{\dim T<j-1} \operatorname{im} D_T$, T a triangulation of a sphere of dimension $<j-1$. The strata at this stage are X_j^T. We then adjoin all D_T with $\dim T = j$ by the appropriate attaching maps on the corresponding S_T and we define $X_{j+1}^T = B_T$ for $\dim T = j$. For $\dim T < j-1$ we let $X_{j+1}^T = X_j^T \cup \bigcup_\sigma \operatorname{im}(c\sigma-*)$ where σ ranges over the simplices of j-spheres $\Sigma \subset S^\infty$ such that $\ell k(\sigma,\Sigma) \in T$.

The final result of this inductive procedure is to have $X^T = \bigcup_{j>\dim T} X_j^T$.

In the interest of clarity, we note that X^T contains B_T as a deformation retract. It is further useful at this point to recall how the cone $c\Sigma$ was decomposed into codimension-0 cells N_*, N_σ (σ a simplex of Σ) in the preliminary to our specification of the decomposition $\{M(X)\}$ of an LS-stratified manifold M. When different spheres are involved we use the notation N_*^Σ, N_σ^Σ etc., and note that the decomposition may be taken to be invariant under A_T, $\Sigma \in T$. Moreover, we must remark that the choice is consistent with the maps $h(\Sigma,\sigma): c\,\ell k(\sigma,\Sigma) \to \sigma^*$. I.e. if we let $\Sigma' = \ell k(\sigma,\Sigma)$ we have

$$h^{-1}N_\sigma^\Sigma = N_*^\Sigma \quad \text{and}$$

$$h^{-1}N_\tau^\Sigma = N_{\tau'}^{\Sigma'} \quad \text{where } \sigma < \tau \text{ and } \tau = \tau' * \sigma$$

with τ a simplex of $\ell k(\sigma(,\Sigma)$. (Of course $h^{-1}N^{\Sigma}_{\tau} = \emptyset$ if $\sigma \not< \tau$.)

Thus the image of $\bigcup_{\Sigma \in T} N^{\Sigma}_{*}$ in B_{LS} is a disc-bundle over B_T. This leads us to another decomposition $\{M_T\}$ of B_{LS}, defined as follows: Let $B^{(j)}$ denote $\bigcup_{\dim T<j} \operatorname{im} D_T \subset B_{LS}$. Assuming that $M_T \cap B^{(j)}$ has already been specified for $\dim T < j-1$ we let $M_T \cap B^{(j+1)} = (M_T \cap B^{(j)}) \cup W$ where $W = \operatorname{image}(\bigcup_{\Sigma_\sigma \in T} N^{\Sigma}_{\sigma})$. For $\dim T = j$ we let $M_T \cap B^{(j+1)} = \bigcup_{\Sigma \in T} \operatorname{im} N^{\Sigma}_{*}$. At the conclusion of this induction we thus have $M_T = \bigcup_j M_T \cap B^{(j)}$. Now set $\hat{X}^T = X^T \cap M_T$. We claim that $B_T \subset \hat{X}^T \subset X^T$ and that the deformation retraction of X^T to B_T may be factored as a deformation retraction of X^T onto $\hat{X}^T$, followed by a deformation retraction of $\hat{X}^T$ onto B_T. Clearly, $B_{LS} = \bigcup_T M_T$. Moreover M_T is a PL disc bundle over $\hat{X}^T$ onto with structure group A_T.

These preliminaries taken care of, we are now ready to indicate how a map $f: M^n \to B_{LS}$, M^n a PL manifold, induces an LS-stratified structure on M^n. First of all we may assume that, modulo deformation, f has the property that $f^{-1}(M_T) = \emptyset$ for $\dim T > n$. This is clear as a matter of elementary general-position considerations. We next consider all T of dimension $n-1$ and make f PL transverse-regular to X^T, so that, in general, $f^{-1}X^T$ is a set of isolated points of M^n. Let Z^T denote this set and note that Z^T has a tubular neighborhood $K_T = f^{-1}M_T$. In fact, $K_T \to M_T$ has the structure of a bundle map over $Z^T \to T^T$. Let $Q_o = \bigcup_{\dim T=n-1} K_T$, and note that $f|Q_o$ is transverse to all X^T for $\dim T$ arbitrary.

Now assume, inductively that for all T of dimension $> n-i$:

a) $f^{-1}(M_T)$ is a codimension-0 submanifold of M^n such that $f|f^{-1}(M_T)$ is transverse to $\hat{X}^T$, and such that $f:f^{-1}(M_T) \to M_T$ is a bundle map covering $f^{-1}(\hat{X}^T) \to \hat{X}^T$.

b) $\bigcup_{\dim T>n-i} f^{-1}(M^T)$ is a codimension-0 submanifold Q_{i-1} of M^n.

It follows that $f|Q_{i-1}$ is transverse to X^T for all T.

Assuming (a) and (b), let T be of dimension $n-i-1$. On ∂Q_i, of course, f is already transverse to X^T. In fact $f^{-1}(X^T) \cap (M^n - \mathrm{int}\, Q_{i-1}) = f^{-1}(\hat{X}^T)$. We then deform f rel Q_{i-1} so as to be transverse to $\hat{X}^T$. In fact $Z^T = f^{-1}(X^T)$ acquires a neighborhood $K_T = f^{-1}(M_T)$ so that $K_T \to M_T$ is a bundle map covering $Z^T \to X^T$. Moreover, we may do this for all $n-i-1$ dimensional T simultaneously, and setting $Q_i = Q_{i-1} \cup \bigcup_{\dim T = n-i-1} K_T$ we have reached a situation where statements (a) and (b) are true with $n-i$ replaced by $n-i-1$. The induction terminates when $i = n$, with $Q_n = M^n$. At this terminal stage, f is simultaneously transverse to all X^T, and is, in fact, LS-stratified, with strata given by the connected components of the various $f^{-1}(X^T)$. If $X \subset M$ is a stratum with $X \subset f^{-1}(X^T)$, then $T = \mathrm{link}\, X$.

Moreover, the decomposition $\{M(X)\}$ of M^n may be specified by taking $M(X)$ to be the component of $f^{-1}(M_T)$ such that X is a component of $f^{-1}(X^T)$ and $M(X) \cap X \neq \emptyset$. Then $X_0 = X \cap M(X)$ is $f^{-1}(\hat{X}^T)$.

It is clear that this argument may be relativized; if $f: M^n \to B_{LS}$ is a map such that $f|\partial M^n$ has all the properties obtained in the terminal stages of the above procedure, then f may be deformed, rel ∂M, so that the resulting map also has these properties. It therefore follows that, up to concordance, the LS-stratified structure on M^n obtained from f depends only on the homotopy class of f in $[M^n, B_{LS}]$. Thus, we have exhibited a map from $[M^n, B_{LS}]$ to the set of concordance classes of LS-stratified structures on M^n.

[We ask the reader to note that the argument above is one to which we shall allude subsequently, in that it is in essence identical to an important step in the proof of subsequent results.]

To conclude the proof of the current lemma, we must show that any LS-stratified structure on M^n arises, up to concordance at least, from a map $M^n \to B_{LS}$, and that concordant LS-stratifications arise from homotopic maps. Let $\{X\}$ be the strata of the stratified manifold M^n and $\{M(X)\}$ the corresponding decomposition of M^n, as earlier described. Now for all 0-strata X, there is clearly a map $f_x : M(X), X \to M_{\mathcal{S}(X)}, B_{\mathcal{S}(X)}$. Moreover, this is a map of stratified spaces in the sense that for any stratum Y with $X < Y$, $Y \cap M(X)$ goes to $X^{\mathcal{S}(Y)} \cap M_{\mathcal{S}(X)}$. Thus, on the subspace $\bigcup_{\dim X=0} M(X)$ of M^n we have a stratum-preserving map $f_0 = \bigcup f_x$.

Suppose, now, inductively that on $M^{(j-1)} = \bigcup_{\dim X<j-1} M(X)$ we have defined a map $f_{j-1}: M^{(j-1)} \to B_{LS}$ such that $f_{j-1}(X) \subset X^{\mathcal{S}(X)}$ and $f_{j-1}(M(X)) \subset M_{\mathcal{S}(X)}$ for all strata X of M^n having dimension $< j-1$. Suppose, moreover, that each map $f_{j-1}|M(X) \to M_{\mathcal{S}(X)}$ is an $A_{\mathcal{S}(X)}$-disc bundle map over $f_{j-1}|X_o \to \hat{X}^{(X)j-1}$. This means in particular, that for every Y of dimension $> j$ with $X < Y$, $f_{j-1}(Y \cap M(X)) \subset X^{\mathcal{S}(Y)} \cap M_{\mathcal{S}(X)}$; furthermore it implies that $f_{j-1}(M(Y) \cap M(X)) \subset M_{\mathcal{S}(Y)} \cap M_{\mathcal{S}(X)}$, and, in fact, that $f_{j-1}|(M(Y)) \cap M(X)) \to M_{\mathcal{S}(Y)}$ is an $A_{\mathcal{S}(Y)}$-disc bundle map over $f_{j-1}|Y_o \cap M(X) \to \hat{X}^{\mathcal{S}(Y)} \cap M_{\mathcal{S}(X)} \subset \hat{X}^{\mathcal{S}(Y)}$.

Therefore, let Y be a j-dimensional stratum and recall the $A_{\mathcal{S}(Y)}$-disc bundle $M(Y)$ over Y_o. Recall too that $\hat{X}^{\mathcal{S}(Y)}$ is of the homotopy type of $B_{\mathcal{S}(Y)}$ and thus $\hat{X}^{\mathcal{S}(Y)}$ is, essentially, the universal classifying space for the group $A_{\mathcal{S}(Y)}$ while $M_{\mathcal{S}(Y)}$ is the associated canonical disc bundle. Thus, $f_{j-1}|Y_o \cap M^{(j-1)} \to \hat{X}^{\mathcal{S}(Y)}$ extends to a map $g_Y: Y_o \to \hat{X}^{\mathcal{S}(Y)}$ while $f_{j-1}|M(Y) \cap M^{(j-1)}$ extends to $f_Y: M(Y) \to M_{\mathcal{S}(Y)}$ with the property that $f_Y|Y_o = g_Y$ and f is an $A_{\mathcal{S}(Y)}$-disc bundle map over g_Y. We thus may define $f_j = f_{j-1} \cup \bigcup_{\dim Y=j} f_Y$ on $M^{(j)} = \bigcup_{\dim X<j} M(X)$ and, clearly, f_j has all the stated properties of f_{j-1} if $j-1$ be replaced by j. Finally, let $f = f_n$. It is clear that

$f: M^n \to B_{LS}$ produces on M^n precisely the LS-stratified structure with which we began, i.e. $f^{-1}(X_{\mathcal{S}(X)}) \supset X$ for all X, f being transverse to all X_T. Thus the map from $[M^n, B_{LS}]$ to classes of LS-stratified structures on M^n is surjective. Furthermore, the argument above is quite easily relativized, so that one may quickly show that if M^n has two LS stratifications M_o, M_1 arising from maps $f_o; f_1: M^n \to B_{LS}$ such that M_o, M_1 are abstractly concordant, then there is a map $F: M \times I \to B_{LS}$ with $F_o = f_o$, $F_1 = f_1$ which induces this concordance. Hence, $[M^n, B_{LS}]$ goes bijectively onto concordance classes of LS-stratified structures on M^n. This completes the proof of 7.4.

7.3. Piecewise-Differentiable Immersions and Their Gauss Maps; The Main Theorem

Our aim in this section is to construct a Gauss map $g(f): M \to \mathcal{G}^c_{n,k}$ where f is a piecewise-differentiable immersion (in a sense soon to be more precisely defined) of the PL manifold M^n in R^{n+k}.

Before discussing the precise class of immersions we have in mind, we must first refine somewhat the notion of stratification introduced in the previous section. In particular, we shall introduce the concept of smooth stratification.

Recall first the definition of strictly stratified manifold from §7.2. Consider the additional requirement that the stratification be smooth in the following sense: Given a strictly-stratified manifold M^n, we shall call it smoothly stratified as well if and only if each stratum is smooth. By this we mean simply that each stratum X may be thought of as a co-dimension-0 submanifold of some smooth manifold X' (without boundary). (X' is merely to be thought of as inducing the smoothness of X. We do not view X' as a subspace of M^n.) Moreover, we require that the strata fit together smoothly in the sense that if $X < Y$, then the inclusion $X \subset Y$ extends to a smooth embedding of a neighborhood of X in X' into Y'.

It is now easy to define a smoothly-stratified manifold in the more general case where the stratification is not strict. In particular, a stratified manifold becomes smoothly stratified when provided with an atlas of charts $\{U_i\}$ such that each U_i is stratified smoothly and strictly, and such that on overlaps of the form $U_i \cap U_j$, the smooth, strict stratifications inherited from U_i and U_j respectively coincide. It follows that the interiors of strata of smoothly stratified manifolds have specific smooth structures.

Let us now narrow the category under consideration to smoothly stratified manifolds whose underlying PL stratification is LS. This is the kind of geometric object whose immersions into Euclidean space we shall be studying.

Accordingly, let us specify precisely what kind of immersion of a smoothly LS-stratified manifold M^n in R^{n+k} generates a Gauss map. First note that given such a manifold, we acquire a certain additional structure on its tangent bundle. Assume, momentarily, that M^n is strictly-stratified. For any stratum X, let X' be the smooth, boundaryless manifold containing X as a codimension-0 submanifold, as posited in the definition of smoothness for stratified manifolds. Let p be an arbitrary point in X. We consider a certain subset of $T_p(X')$: Let ρ be a smooth trajectory $\rho: R \to X'$ with $\rho(0) = p$, and $\rho([0,\varepsilon]) \subset X$ for some small $\varepsilon > 0$. The 1-jet $[\rho]$ of ρ at 0 is, essentially, a tangent vector to X' at p. We let $T_p(X)$ be the union of all such vectors. In general, $T_p(X)$ is a Euclidean cone but not necessarily a subvector-space of $T_p(X_1)$.

We may then view $T_p(M^n)$ as $\bigcup_{p\varepsilon X} T_p(X)$. The topology arises from noting that if $p \varepsilon X_1 < X_2$ then $T_p(X_1)$ is naturally viewed as a sub-cone of $T_p(X_2)$, independent of the particular choice of X_1' and X_2'. Thus, the union must respect this identification, and takes on the obvious weak topology of union.

Now we drop the assumption that M^n be strictly stratified and note that the decomposition of $T_p(M^n)$ as $\bigcup_{p\varepsilon X} T_p(X)$ may still be made since a stratum X containing p is, locally, the union of strata of a strictly stratified neighborhood of p.

We are now in a position to characterize the immersions that we really want to consider. First of all, we demand that an immersion $f: M^n \to R^{n+k}$ be smooth on each stratum. (For a strictly-stratified manifold, this means that for each stratum X, $f|X$ extends to a smooth map on X'. In the case of non-strictly-stratified M^n, of course, we merely impose the condition above on each strictly-stratified chart.)

This condition, of itself, is not enough for our purposes. Note that, in its presence, we obtain a well-defined "differential" $df_p : T_pM^n \to T_{f(p)}(R^{n+k})$. To see how this is defined assume, for the moment, that M^n is strictly-stratified; then we may define, for $p \in X$, df_p on $T_p(X)$ as $df_1|T_p(X)$ where f_1 is a smooth extension of $f|X$ to a neighborhood of X in X'. Clearly, df_p is well-defined in this way, i.e. it does not depend on which $T_p(X)$ we regard as containing a specific element of $T_p(M)$. Equally clearly, the definition of df_p may be made to hold for non-strictly-stratified manifolds as well.

Of course, we do not claim that $df = \bigcup_{p \in M} df_p$ is a continuous map on the tangent bundle of M^n. In fact, there is no "natural" topology on $\bigcup_p T_p(M^n)$, i.e. no canonical way of identifying it with the PL tangent bundle of the underlying PL manifold M^n. However when we restrict to the interior $\mathring{X}$ of a single stratum X (i.e. $\mathring{X} = X - \bigcup_{Y<X} Y$) we find that $T_X(M^n) = \bigcup_{p \in \mathring{X}} T_p(M^n)$ does have a natural topology, with respect to which $df|T_XM^n \to TR^{n+k}$ <u>is</u> continuous.

Now in the case of an immersion f of a smoothly LS-stratified M^n, f smooth on each stratum as above, we may establish the further requirement that $df_p : T_p(M^n) \to T_{f(p)}(R^{n+k})$ be a homeomorphism onto its image. Moreover, we note that if this requirement is fulfilled, then the cone $df_p(T_p(M^n)) \subset T_{f(p)}(R^{n+k})$ may be represented as a certain direct sum, i.e. $df_p(T_pM^n)) = P \oplus Q$ where P is the sub-vector-space $d(f|\mathring{X})(T_p(X))$ where $p \in \mathring{X}$, and where $Q = df_p(T_p(M^n)) \cap P^{\perp}$. This fact is easily seen since the cone $T_p(M^n)$ may be decomposed as $T_p(X) \oplus Q'$ for some cone Q'. Since df_p is a map of cones, it follows that Q may be taken to be the image of $df_p(Q)$ under projection $T_{f(p)}(R^{n+k}) \to P^{\perp}$.

Now Q', and hence Q, are naturally associated with a certain combinatorial object. That is, Q' may be thought of as the infinite cone $c(\mathring{X})$ on the simplicial complex $\mathcal{L}(X)$. The same picture holds for Q, therefore. That is, if we let $S(U_p)$ be the unit sphere in the sub-space $U_p = P^{\perp} \subset T_{f(p)}(R^{n+k})$ then $Q \cap S(U_p) = \Sigma$ is a topological sphere, and it is stratified, moreover, where the strata are of the form $df_p(T_pY) \cap S(U_p)$, Y ranging over the strata such that $X < Y$. Our last requirement of df is that, for each p, Σ is admissibly triangulated with simplices $\sigma(Y) = df_p(T_pY) \cap S(U_p)$ for each Y with $X < Y$. Here $\dim \sigma(Y) = \dim Y - \dim X - 1$.

To summarize, then: We shall be interested in smoothly LS-stratified manifolds M^n and immersions $f: M^n \to R^{n+k}$ such that

α) f is piecewise smooth, i.e. smooth on each stratum (of any strictly-stratified local chart).

β) df is nowhere singular, i.e. for all p, $df_p: T_p(M^n) \to T_{f(p)}(R^{n+k}_\circ)$ is a homeomorphism onto its image.

γ) for any $p \in X$ (X a stratum of dimension $n-j$), the natural stratification of $df_p(T_pM)$ is of the form $df_p(T_p\dot{X}) \oplus c\Sigma_p$ where Σ_p is an admissibly-triangulated $(j-1)$-sphere in the unit sphere of $[df(T_pX)]$, Σ_p being simplicially equivalent to $\mathcal{L}(\dot{X})$.

Now note that when f is an immersion as above, and $p \in X$, where X is an $(n-j)$-dimensional stratum, we obtain a certain formal link of dimension j. That is, with $U_p = [df(T_pX)]$, we have: $L(p,f) = (U_p, \Sigma_p)$. (U_p is identified with a vector subspace of R^{n+k} in the obvious way.)

We now assert that given a smoothly LS-stratified manifold M^n and sufficiently large k, there exists an immersion (in fact an embedding) $M^n \to R^{n+k}$ satisfying α, β, γ. Moreover, this will be unique up to isotopy. We leave this to the reader.

The Gauss map $g(f): M^n \to \mathcal{G}^c_{n,k}$ of an immersion $f: M^n \to R^{n+k}$ satisfying α, β, γ may now be defined. We note in advance that, in a rather unimportant way, $g(f)$ fails to be canonical if we take canonical to mean that $g(f)$ is determined exactly on each point of

M^n by the geometric data of the immersion. However, up to an isotopy of M^n, the Gauss map will be well-defined so that this failure is of no more than passing curiosity.

Given $f: M^n \to R^{n+k}$, recall the decomposition $\{M(X)\}$ of M^n. We begin be defining $g(f)$ on $M(X)$ for all 0-dimensional strata X. This is easily done since $M(X)$ is identified with N^{Σ}_* in a specific way, where Σ is the link $\mathcal{L}(X)$. At the same time, we have a certain formal link $L(x,f)$ of dimension n, where $\{x\} = X$, and $\Sigma_{L(x,f)} = \Sigma(x,f)$ is isomorphic, in a specific way, with Σ. But then, since there is a standard way of identifying N^{Σ}_* with $c\Sigma$, we may identify $M(X)$ with $c\Sigma(X,f)$ and thus map $M(X)$ to $\mathcal{G}^c_{n,k}$ by composing this identification map with $c\Sigma(x,f) \to e_{L(x,f)} \subset \mathcal{G}^c_{n,k}$. Doing this for all 0-dimensional X yields the definition of $g(f)$ on $M^{(0)} = \bigcup_{\dim X=0} M(X)$.

Suppose now, inductively, that $g(f)$ has been defined on $M^{(j-1)} = \bigcup_{\dim X<j} M(X)$. We wish to define $g(f)$ on $M(X)$ for all X of dimension j. Recall that $M(X)$ is parameterized as a twisted product $X_0 \times_\tau N^{\Sigma}_*$, where $\Sigma = \mathcal{L}(X)$. Again, we may re-parameterize N^{Σ}_* as $c\Sigma$ and thus we have a way of thinking of $M(X)$ as $X_0 \times_\tau c\Sigma$. Moreover, for any $x \in X$, we may identify $x \times c\Sigma \subset X \times_\tau c\Sigma$ with $c\Sigma(x,f)$. Hence, we have a continuous map $\operatorname{int} X_0 \to \mathcal{L}_\Sigma$ given by $x \to L(u(x),f)$, where u is the map $X_0 \to X$ described earlier when the decomposition $\{M(X)\}$ was specified. Now it is straightforward to cover $u|\operatorname{int} X_0$ by a bundle map from the normal bundle of $\operatorname{int} X_0$ in M to the normal bundle of $\dot{X}$. Consequently, the map $x \to L(u(x),f)$ of X_0 into $\mathcal{L}_\Sigma$ may be covered by a bundle map $(\operatorname{int} X_0) \times_\tau c\Sigma \to P_\Sigma$ where P_Σ is the natural $c\Sigma$ bundle over $\mathcal{L}_\Sigma$ described in 7.1. But P_Σ is the pre-image of $N_\Sigma \subset \mathcal{G}^c_{n,k}$ and thus we receive a map g_X from $\operatorname{int} X_0 \times_\tau c\Sigma \subset M(X)$ to N_Σ. We have already, by assumption, defined $g(f)$ on the remainder of $M(X)$

(i.e. $M(X)-M^{(j-1)} = (\text{int } X_o \times_\tau c\Sigma)$). We claim that g_X is consistent with $g(f)|M^{(j-1)}$ i.e. $g(f)|M^{(j-1)} \cup g_X$ is a continuous map on $M^{(j-1)} \cup M(X)$. In other words, we may define $g(f)$ on $M^{(j)} = M^{(j-1)} \cup \bigcup_{\dim X=j} M(X)$ by $g(f)|M^{j-1} \cup \bigcup_{\dim X=j} g_X$.

We shall leave to the reader the verification of the claim of continuity.

Note that the construction of $g(f)$ depends on the choice of the decomposition $\{M(X)\}$ and on the parameterization of each $M(X)$. However, since the choices are unique up to a stratum-preserving isotopy of M^n, the Gauss map is unique in precisely the same sense.

Of course, we are also obliged to define the bundle map $TM^n \to \gamma^c_{n,k}$ which covers the Gauss map $g(f)$ of an immersion. It will suffice to characterize this bundle map on $TM|M(X)-M^{(j-1)}$ for all j-strata X, all j. Once more, parameterize $M(X)$ as $X_o \times_\tau c\Sigma$, $\Sigma = \mathcal{L}(X)$. We see that $TM|M(X)$ is naturally isomorphic to $(u\circ\pi)^*TX \oplus \pi^*\eta$ where η is the PL $(n-j)$-bundle $X_o \times_\tau c\Sigma$ whose projection is π. More specifically, let $x \in \text{int } X_o$ and consider $x \times c\Sigma \subset M(X)-M^{(j-1)}$. In a natural way $TM|C = \tau \oplus \xi$ where ξ is the trivial bundle with constant fiber $T_{u(x)}X$. Here we are using u to identify T_xX_o with $T_{u(x)}$. Now let D denote the image fiber disc of C in P_Σ, i.e. $D = c\Sigma_{L(u(x),f)}$. Then $\gamma^c_{n,k}$ pulled back to D is merely $TD \oplus \theta$ where θ is the trivial bundle with constant fiber $(U_{L(u(x),f)})^\perp = (U(u(x),f))^\perp$. Since, on C, $g(f)$ factors through a homeomorphism $C \to D$ and $df|T_{u(x)}X$ identified $T_{u(x)}X$ with $U(u(x),f))^\perp$, we have a natural identification of $TM|C$ with the pullback to C, via D, of $\gamma^c_{n,k}$. Thus the covering of $g(f)$, as it restricts to C, is defined, and since every point of M is in precisely one such C, we get a bundle map $TM \to \gamma^c_{n,k}$ covering $g(f)$. (Again, verification that this map is, first of all, continuous, and, in addition, a bundle map is left to the reader.)

We now specify what is to be meant by a "geometric subspace" of $\mathcal{G}^c_{n,k}$, i.e., those spaces which are to play a role analogous to that of geometric subcomplexes of $\mathcal{G}_{n,k}$. In this regard, recall that if T is an equivalence class of triangulations of S^{j-1}, then $\mathcal{L}_T$ = the set of formal j-dimensional links $L = (U,\Sigma)$ with $\Sigma \in T$ admits a natural map to the standard Grassmannian, $\theta_T: \mathcal{L}_T \to G_{n-j,j+k}$ defined by $(U,\Sigma) \to U$; by 7.1, this is a fibration. Recall also that we think of $\mathcal{L}_T$ as embedded in $\mathcal{G}^c_{n,k}$ with a certain neighborhood N_T.

Consider subspaces $H \subset \mathcal{G}^c_{n,k}$ with the following property

(1) If $H \cap (\text{int } N_T) \neq \emptyset$, then $H \cap (\text{int } N_T)$ is an open disc bundle (i.e. with fiber $cT-T$) over $H \cap \mathcal{L}_T$. Moreover $H \cap \mathcal{L}_T$ is open in $\mathcal{L}_T$ and $\theta_T(H \cap \mathcal{L}_T)$ is open in $G_{n-j,j+k}$. Finally, $\theta_T | H \cap \mathcal{L}_T$ is a fibration.

We call such subspaces "geometric." The obvious examples of such subspaces are of the form $\bigcup_{T \in \mathcal{V}} N_T$ where $\mathcal{V}$ is some collection of equivalence-classes of triangulations having the property that if $\Sigma \in T \in \mathcal{V}$ and v is a vertex of Σ, then $[\ell k(v,\Sigma)] \in \mathcal{V}$.

The main theorem of this section may now be stated:

<u>7.5 Theorem</u>. Let H be a geometric subspace of $\mathcal{G}^c_{n,k}$. Let M^n be a PL manifold with no closed components. Suppose that the map $h: M^n \to H$ is covered by a bundle map $TM^n \to \gamma^c_{n,k}|H$. Then:

M^n admits a stratification as an LS-stratified manifold, so that, with respect to this stratification, there is a piecewise-differentiable immersion $f: M^n \to R^{n+k}$ (satisfying α, β, γ) so $g(f): M^n \to \mathcal{G}^c_{n,k}$ has its image in H and, moreover, $g(f)$ is homotopic to h in H.

There is, as well, a relative version of this result:

<u>7.6 Corollary</u>. Let V^n be a smoothly LS-stratified manifold and let M^n be obtained from V^n by adding handles of dimension $< n$. Suppose $f_o|V^n \to R^{n+k}$ is a piecewise-differentiable immersion, with

respect to this stratification so that $g(f_o)\colon V^n \to H \subset \mathcal{G}^c_{n,k}$ (H geometric) extends to $h\colon M \to H$ while the bundle map $TV^n \to \gamma^c_{n,k}|H$ covering $g(f_o)$ extends to a bundle map $h\colon TM^n \to \gamma^c_{n,k}|H$ covering h. Then:

There is a smooth LS-stratification of M^n extending that of V^n and an immersion $f\colon M^n \to R^{n+k}$ extending f_o, and f piecewise differentiable with respect to this stratification. Moreover, $g(f)\colon M^n \to \mathcal{G}^c_{n,k}$ has its image in H and g(f) is homotopic to h, rel $g(f_o)$, in H.

We shall briefly postpone the proof of 7.5, in order to establish some consequences first. We take note of the fact that 7.5 is, in fact, somewhat stronger than the analogous result 4.2 relating piecewise-linear immersions to the geometric subcomplexes of $\mathcal{G}_{n,k}$. We may use this additional strength to good effect in analyzing the homotopy type of $\mathcal{G}^c_{n,k}$. Note, first of all, that the double sequence

$$\begin{array}{ccccccc} & & \vdots & & \vdots & & \\ \cdots & & \mathcal{G}_{n,k} & \xrightarrow{\alpha} & \mathcal{G}_{n+1,k} & & \cdots \\ & & \downarrow{\scriptstyle\beta} & & \downarrow{\scriptstyle\beta} & & \\ \cdots & & \mathcal{G}_{n,k+1} & \xrightarrow{\alpha} & \mathcal{G}_{n+1,k+1} & & \cdots \\ & & \vdots & & \vdots & & \end{array}$$

passes over, under retopologization, to a similar double sequence

$$\begin{array}{ccccc}
 & \vdots & & \vdots & \\
\cdots & \mathcal{G}^c_{n,k} & \xrightarrow{\alpha} & \mathcal{G}^c_{n+1,k} & \cdots \\
 & \downarrow \beta & & \downarrow \beta & \\
\cdots & \mathcal{G}^c_{n,k+1} & \xrightarrow{\alpha} & \mathcal{G}^c_{n+1,k+1} & \cdots \\
 & \vdots & & \vdots &
\end{array}$$

which maps to BPL by the map which, at each stage, classifies the stabilization of $\gamma^c_{n,k}$. This suggests the conjecture that $\lim_{n,k} \mathcal{G}^c_{n,k} \to BPL$ is a homotopy equivalence, which conjecture, however, is not quite true. There is a more or less obvious obstruction to the truth of the conjecture which, moreover, is the only one. Informally, this obstruction arises as follows: Consider LS-stratified manifolds, ignoring for the moment the smoothability of the stratification. More specifically consider an LS-stratified j-sphere; we ask whether this sphere is the boundary of an LS-stratified (j+1)-disc. Obviously the answer is "yes" when, e.g., the stratification of the sphere is a combinatorial triangulation, for then we merely triangulate the (j+1)-disc as the simplicial cone on the j-sphere. However, it is clear that this construction will not work for more general LS-stratifications of the sphere, for then the cone will be stratified, but the 0-stratum at the cone point will have for its link the original j-sphere, which is not, in this case, simplicial. Hence the stratification of the disc fails to be LS.

We may, however, note the following about the natural map $\phi: \mathcal{G}^c = \lim_{n,k} \mathcal{G}^c_{n,k} \to BPL$.

<u>7.7 Lemma</u>. For any finite complex K, $[K,\mathcal{G}^c] \xrightarrow{\phi_*} [K,BPL]$ is surjective; moreover, there is a natural splitting $t: [K,BPL] \to [K,\mathcal{G}^c]$

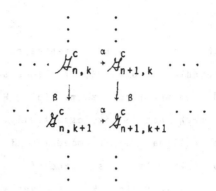

of $\phi*$.

Proof: Given the map $a: K \to BPL$, represent its homotopy class by the PL n-plane bundle α over K, for some sufficiently large n. Indeed, by [Wa], we may find some manifold W^n of the simple homotopy type of K such that α is, essentially TW^n. Then, for large enough k, we shall have a PL-embedding $f: W^n \to R^{n+k}$, that is, W^n is a piecewise-linear subspace of R^{n+k}, with respect to some triangulation. Ipso facto, W^n is given a smooth LS stratified structure for which the embedding f is piecewise-differentiable and obviously $g(f): W^n \to \mathcal{G}^c_{n,k}$ has the property that $\phi \circ g(f)$ classifies α. Hence the surjectivity claimed by the Lemma is established. But, in addition, the manifold W^n, the triangulation thereof, and the PL embedding are uniquely determined up to concordance (if n,k be large enough). Hence this construction produces a unique element if $[K, \mathcal{G}^c_{n,k}]$ and hence in $[K, \mathcal{G}^c]$ as the pre-image of $[a]$. Thus the construction defines the splitting map t as required.

We may say something more about $\phi: \mathcal{G}^c \to BPL$, at least on the homotopy group level. That is, we shall characterize geometrically the kernel θ_i of $\phi*: \pi_i\mathcal{G}^c \to \pi_i BPL$. Of course $\pi_i\mathcal{G}^c$ splits as $\pi_i BPL \oplus \theta_i$. (In fact, it is possible to describe a space B_{SLS} with $\theta_i = \pi_i B_{SLS}$ but this would take us too far afield.)

Consider all smooth LS-stratifications of S^i, and call two such equivalent if they are concordant, i.e. S^i_o and S^i_1 are equivalent when there is a smooth LS stratification T of $S^i \times I$ whose boundary is $S^i_o \amalg S^i_1$. Let θ_i denote the set of equivalence classes. Clearly, there is a distinguished element of θ_i, namely that represented by the stratification with only one stratum, i.e. all of S^i (with its usual smooth structure). It is easily seen that a stratification of S^i falls within this class if and only if it is the boundary of a smoothly LS-stratified $(i+1)$-disc.

θ_i is, in fact, a group, as will be seen shortly. First, however, note that there is a natural map $\theta_i \xrightarrow{\psi} \pi_i \mathcal{G}^c$. We omit some technicalities concerning the role of the base point, and simply define ψ by noting that given an element in θ_i, we may represent it by a smoothly LS-stratified i-sphere which, for clarity, we denote T, and then, we claim, proceed to embed T in some Euclidean space R^{i+k}, the embedding f being piecewise differentiable and satisfying conditions α, β, γ. We thus obtain a Gauss map $g(f): T \cong S^i \to \mathcal{G}^c_{i,k}$, which we may consider to represent an element of $\pi_i \mathcal{G}^c$. We claim that this element depends only on the equivalence class of T, and not on T itself, nor the embedding f.

<u>7.8 Lemma</u>. $0 \to \theta_i \xrightarrow{\psi} \pi_i \mathcal{G}^c \xrightarrow{\phi*} \pi_i BPL \to 0$ is exact.

We have already seen, in essence, that $\phi*$ in the above sequence is split surjective. We note that $\text{im}\ \psi \subset \ker \phi*$ merely because, for the Gauss map of any piecewise differentiable immersion $f: M \to R^{n+k}$, $\phi \circ g(f)$ classifies the stable tangent bundle of M; thus if $M = S^i$, $\phi \circ g(f)$ is homotopically trivial, hence $\phi^* \circ \psi$ is trivial.

We shall define a map ω: $\ker \phi* \to \theta_i$ which is inverse to ψ. Suppose $u: S^i \to \mathcal{G}^c$ with $\phi*[u] = 0$ in $\pi_i BPL$. Think of u as a map $S^i \to \mathcal{G}^c_{n,k}$, n and k both large. We may then replace S^i by $W = S^i \times D^{n-i}$ and think of u as a map $W \to \mathcal{G}^c_{n,k}$ covered by a particular bundle map of the (trivial) tangent bundle of W to $\gamma^c_{n,k}$. If we now apply the main result 7.5, we may immerse (in fact, embed if k be large enough) W^n in R^{n+k} via f so that, with respect to some smooth LS-stratification of W, f has properties α, β, γ and the Gauss map g(f) is homotopic to u.

Now think of S^i as $S^i \times \{0\} \subset S^i \times D^{n-i} = W \subset R^{n+k}$. After a small isotopy, we may assume that S^i meets the strata of W transversally. I.e., S^i is stratified so that a typical codimension-j stratum Y is a component of $S^i \cap X$ for some codimension j stratum X of W. Moreover, since X is smooth and Y has a trivial normal bundle in X, it follows that Y must inherit smoothability. More precisely, the embedding of W in R^{n+k} is concordant to one in which $S^i \subset W$ is a "submanifold" of the smooth LS-stratified manifold W. This stratification of S^i defines an element of θ_i which we denote by $\omega([u])$. We claim that $\omega([u])$ is well-defined, details being left to the reader with the advice that the relative version 7.6 of the main theorem is a key ingredient. So, too, is the further fact that the standard maps $\alpha: \mathcal{G}^c_{n,k} \to \mathcal{G}^c_{n+1,k}$ and $\beta: \mathcal{G}^c_{n,k} \to \mathcal{G}^c_{n,k+1}$ are well-behaved with respect to Gauss maps in the sense that if $f: M^n \to R^{n+1}$ is a piecewise-smooth immersion then $\alpha \circ g(f) \sim g(f \times id)$ where $f \times id: M^n \times R \to R^{n+k} \times R = R^{n+k+1}$, while $\beta \circ g(f) = g(f')$ where f' is the composition $M^n \xrightarrow{f} R^{n+k} \subset R^{n+k+1}$.

It is clear that $\omega \circ \psi$ is the identity on θ_i. We must show that ω is injective. Consider once more the situation outlined above, i.e. S^i sitting in $W \cong S^i \times D^{n-i}$ as a submanifold (in the sense of smoothly LS stratified manifolds.) Note that S^i has a neighborhood of the form $W' = S^i \times D'$ where D' is a smaller copy of D^{n-i} and where the smooth LS stratification on W' inherited from W is the product of the stratification on S^i with a trivial stratification of D'. It is clear that, regarded as an element of $\pi_i \mathcal{G}^c_{n,k}$, $g(f) \sim g(f|W')$. Now deform $f|W' = f'$, through piecewise smooth embeddings, to $f_1: W' \to R^{n+k}$ so that f_1 is of the form $e \times i$ where $e: S^i \to R^{i+k}$ and i is the standard injection $D' \cong D^{n-i} \subset R^{n-i}$. We leave it to the reader to verify that such a deformation is possible when k is large. It follows that $g(f_1) \sim g(f)$. But note that $g(f_1)|S^i = \alpha^{n-i} g(e)$ where α^{n-i} is the iteration of standard maps, $\alpha^{n-i}: \mathcal{G}^c_{i,k} \to \mathcal{G}^c_{n,k}$. Also recall that $g(e)$, hence $\alpha^{n-i} \circ g(e)$ represents $\psi(\omega([u]))$. But $g(e)$

represents [u] hence $\psi\circ\omega = id$. This completes the proof.

7.4 Proof of Theorem 7.5

As the first step in proving 7.5 we must analyze the geometric structure of a typical geometric subspace H of $\mathcal{G}^c_{n,k}$. In particular, one sees that H has a stratified structure essentially analogous to that of the space B_{LS} of §8.2.

Given $H \subset \mathcal{G}^c_{n,k}$, let T be an equivalence class of triangulations of the i-sphere, $-1 < i < n-1$, so that $\mathcal{L}_T \cap H$ is non-void. We shall use the notation $\mathcal{L}_T \cap H = \bar{\mathcal{L}}_T$. Recall that $N_T \subset \mathcal{G}^c_{n,k}$ is the image of a disc bundle P_T over $\mathcal{L}_T$; therefore, we use $\bar{N}_T$ to denote the image of $P_T|\bar{\mathcal{L}}_T$, recalling that, by the definition of geometric subspaces $\text{int}\, N_T \cap H = \text{int}\, \bar{N}_T$. We now use an inductive procedure to build up a stratification of H having one stratum X^r_T for each component $\bar{\mathcal{L}}^r_T$ of $\bar{\mathcal{L}}_T$. Let $A_0 \subset \ldots \subset A_i \subset A_{i+1} \subset \ldots \subset A_n = H$ be a filtration of H defined by $A_i = \bigcup_{\dim T < i=1} \bar{N}_T$. (Thus, in particular A_0 is $H \cap \mathcal{L}_0 = H \cap G_{n,k}$.) On A_i we define a stratification with open strata $X^{r,i}_T$, one for each component $\bar{\mathcal{L}}^r_T$ with $\dim T < i-1$. We shall have $\overset{\circ}{X}{}^{r,i}_T \subset \overset{\circ}{X}{}^{r,i+1}_T \subset \ldots \subset \overset{\circ}{X}{}^{r,n}_T$ and $\overset{\circ}{X}{}^{r,n}_T = \overset{\circ}{X}{}^{r}_T$. X^r_T is the closure of $\overset{\circ}{X}{}^{r}_T$.

At the 0th stage, let $\overset{\circ}{X}{}^{r,o}_o$ simply be the components of $\bar{\mathcal{L}}_o$.

Suppose, now, inductively, that we have stratified A_{i-1}, with one stratum $\overset{\circ}{X}{}^{r,i-1}_T$ for each component $\bar{\mathcal{L}}^r_T$ of $\bar{\mathcal{L}}_T$, $\dim T < i-2$. Consider any pair (L,σ) where $L = (U_L, \Sigma_L)$ is an i-dimensional formal link such that $[\Sigma_L] = S$ with $L \in \bar{\mathcal{L}}_S \neq 0$, and where σ is a simplex of Σ. Then e_L, the image of $c\Sigma_L$ lies in $\bar{N}_S \subset H$. Furthermore $L_\sigma \in \bar{\mathcal{L}}^r_T \subset \bar{\mathcal{L}}_T$ for $T = [\Sigma_{L_\sigma}]$, and some component $\bar{\mathcal{L}}^r_T$ of $\bar{\mathcal{L}}_T$. Keep in mind the subspace $\text{im}(\overset{\sigma}{c}\sigma\text{-}*) \subset \text{im}\, c\Sigma_L \subset \bar{N}_S$. Let $Z^r_T = \bigcup \text{im}(\psi\sigma\text{-}*)$ where σ ranges over all such σ as above i.e. σ a simplex of Σ_L, $L \in \bar{\mathcal{L}}_S$ for some S and $L_\sigma \in \bar{\mathcal{L}}^r_T$ for the given T

and r. Let $\mathring{X}^{r,i}_T = \mathring{X}^{r,i-1}_T \cup Z^r_T$. This takes care of all strata of A_i corresponding to T of dimension $< i-2$. For T of dimension $i-1$, we simply let $\mathring{X}^{r,i}_T$ be the corresponding component $\bar{\mathcal{Z}}^r_T$ of $\bar{\mathcal{Z}}_T \subset A_i - A_{i-1}$.

Recall the hypothesis of 7.5; we have a non-closed manifold M^n, a map $h\colon M^n \to H$ covered by a bundle map $TM^n \to \gamma^c_{n,k}|H$. We may now define h to make it transverse to the stratification of H. That is, just as in the proof of Lemma 7.4 of §7.2, after suitable deformation, h may be taken to be a stratum-preserving map from an LS-stratification on M^n into H (the stratification of H being the one just described). Thus, if X^r_T is a stratum of H, then $h^{-1}(X^r_T)$ has, as its components, strata of M^n whose link is T. Of course, we may still assume that h is covered by a bundle map $TM^n \to \gamma^c_{n,k}|H$ since the deformation may be covered by a deformation of bundle maps. Henceforth, we shall regard this stratification of M^n as fixed.

7.9 Lemma. The stratification of M^n arising as above can be smoothed in a natural way.

Proof: Assume for the moment that M^n is in fact strictly stratified so that a typical stratum Y is a submanifold of M^n. Let T be the link of $\mathring{Y}$ so $h|Y \to X^r_T$ for some r. We shall smooth the interior Y of Y, but this is essentially equivalent to smoothing Y itself. Note that the open stratum $\mathring{X}^r_T$ contains $\bar{\mathcal{Z}}^r_T$ as a deformation retract (this may be seen directly from the construction). We note, leaving details to the reader, that $\mathring{X}^r_T$ has an open neighborhood in H which may be identified with the open-PL-disc bundle over $\mathring{X}^r_T$ associated to the pullback to $\mathring{X}^r_T$ (under the deformation retraction) of the PL bundle $P_T|\bar{\mathcal{Z}}^r_T \to \bar{\mathcal{Z}}^r_T$. By way of clarification, it should be understood that over $\bar{\mathcal{Z}}^r_T$

itself, the neighborhood looks like $\operatorname{int} \overline{N}^r_T$. For notational simplicity, let the bundle $P_T|\overline{\mathcal{Z}}^r_T$ be denoted by ζ^r_T and the retraction by $\rho^r_T: \overset{\circ}{X}{}^r_T \to \overline{\mathcal{Z}}^r_T$. Let $\nu(Y,M)$ denote the normal bundle of $\overset{\circ}{Y}$ in M^n. Then, by the "transversality" of h, we have:

$$\nu(\overset{\circ}{Y},M) = h^*\rho^{r*}_T \zeta^r_T$$

But recall, in light of §7.1, that $\gamma^c_{n,k}|\overline{\mathcal{Z}}^r_T = \zeta^r_T \oplus \eta^r_T$ where η^r_T is a certain vector bundle canonically induced from the map $\theta_T|\overline{\mathcal{Z}}^r_T \to G_{n-j,k+j}$ (dim T = j-1). Thus we have: $TM^n|\overset{\circ}{Y} = h^*\rho^{r*}_T(\zeta^r_T \oplus \eta^r_T)$. Hence, at least stably, we have, by cancellation: $T\overset{\circ}{Y} =_s h^*\rho^{r*}_T \eta^r_T$. I.e. the PL tangent bundle of Y reduces to a vector bundle and, by classical smoothing theory (see, e.g. [Hi-M]) $\overset{\circ}{Y}$ (and thus Y) is smoothable, and, in fact, endowed with a specific smoothing. (That is, the stable bundle map $[T\overset{\circ}{Y}]_s \to [\eta^r_T]_s$ may be chosen so that its sum with $\nu(\overset{\circ}{Y},M) \to \zeta^r_T$ is equivalent to the given $TM|\overset{\circ}{Y} \to \eta^r_T \oplus \zeta^r_T$.)

It is not hard to show that the reduction of tangent bundles of strata as obtained above is consistent with incidence relations. That is, if $Y^p_1 < Y^q_2$ are incident strata, then there is a canonical way of identifying $T\overset{\circ}{Y}_2|\overset{\circ}{Y}_1$ with $T\overset{\circ}{Y}_1 \oplus \varepsilon^{q-p}$. With regard to this identification, the stable reduction of $T\overset{\circ}{Y}_2$ to a vector bundle as obtained above induces a reduction of $T\overset{\circ}{Y}_1$. We claim that this is identical to the reduction of $T\overset{\circ}{Y}_1$ that has already been produced. It follows, once more by classical smoothing theory, that all the strata may be simultaneously smoothed so that, with $Y_1 < Y_2$, $Y_1 \cap Y_2$ is smooth, and thus the stratification is smooth as well.

We may now remove the ad hoc assumption that M^n was strictly-stratified by dint of the usual observation that a stratified manifold is, locally, strictly-stratified. That is, we may smooth strictly stratified co-ordinate patches by the method outlined above so that, on overlaps, the smoothness structures of the respective strata coincide.

We note, however, that for our purposes we shall, in fact, need slightly stronger properties for the map h than have already been shown. We characterize this as follows: Recall that for an open stratum $\overset{\circ}{X}{}^r_T$ of H, $\gamma^c_{n,k}|X^r_T = \rho^{r*}_T \zeta^r_T \oplus \rho^{r*}_T \eta^r_T$. Designate these summands by ζ' and η' respectively. The property we wish to have, then, is: If Y is a stratum of M^n with link $Y = T$ and $h(Y) \subset X^r_T$ then the bundle map $TM^n|\overset{\circ}{Y} \to \gamma^c_{n,k}|\overset{\circ}{X}{}^r_T$ covering $h|\overset{\circ}{Y}$ splits into a direct sum of bundle maps:

i) $z: \nu(\overset{\circ}{Y},M) \to \zeta'$

ii) $y: T(\overset{\circ}{Y}) \to \eta'$

where z is the natural map arising from the fact that h is transverse to the stratification of H. (We note that the previous argument on the smoothability of Y does not quite yield this property for h, inasmuch as we merely obtained a map from the stabilization of $T\overset{\circ}{Y}$ to the stabilization of η'.)

Let us denote the property defined above as property A.

We wish to show that under the hypotheses of 7.5, h may be assumed to have property A. This demonstration will call upon one of the hypotheses of 7.5 which has heretofore been ignored, namely, the assumption that M^n has no closed components.

<u>7.10 Lemma</u>. Suppose $h: M^n \to H$ is a map transverse to the stratification of H and is covered by a PL bundle map $TM^n \to \gamma^c_{n,k}|H$. Let Y be a stratum of the induced LS stratification of M^n, and let hd Y denote the homotopy dimension of Y modulo the incident strata to Y. Suppose that for all Y of dim > 0, hd $Y <$ dim Y.

Then the bundle map deforms to one having the property A.

The proof of the lemma is quite straightforward. We advert to the argument of 7.9 which showed that a vector bundle structure could be imposed on the stabilization of $T\overset{\circ}{Y}$. We obtained a stable map $(T\overset{\circ}{Y})_s \to \eta'_s$ where the s in subscript denotes stabilization. However, if hd $Y <$ dim Y = dim η', then this stable map desuspends to an unstable map by standard obstruction theory, and the vector bundle structure on $T\overset{\circ}{Y}$ obtained in this way may be identified with the vector-bundle structure on $T\overset{\circ}{Y}$ arising from the smoothing. Thus 7.10 is proved.

We return to the proof of 7.5. Our assumption thus far is that h is transverse to the stratification of H, in which case the induced stratification of M^n has been shown, via 7.9, to be smoothable. However, the hypothesis of 7.10 has not been demonstrated. We achieve this as follows: Suppose M^n (or a component thereof) is compact and has a non-void boundary. For each stratum Y of M^n with $hd\ Y = \dim Y$, remove an open n-disc of the form of a small tubular neighborhood of a small open disc of dimension = $\dim Y$ in $\operatorname{int} Y$. (All the removed discs are disjoint.) Order these discs and connect the boundary of the ith to the boundary of the $(i+1)$st by an arc, such that each arc is transverse to all strata (i.e. it meets only strata of dimension n and $n+1$, the former in subarcs, the latter in discrete points). We may assume the arcs are disjoint (by general position if $n > 3$; with some slight additional argument if $n = 2$, where some care may be needed in the ordering). Remove small tubular neighborhoods of these arcs. The effect thus far is to puncture M^n once. Now connect the boundary of what has been removed to ∂M^n by an arc and remove a tubular neighborhood of the arc. Thus we obtain M', a smaller copy of M^n, with an inherited stratification and h may be replaced by $h' = h|M'$, which does satisfy the hypothesis of 7.9.

If M^n has void boundary or is non-compact, a slightly more elaborate argument may be made along the same lines; this is left to the reader as an exercise.

Therefore, it may now be assumed, in the proof of 7.5, that the map h satisfies property A. In view of the argument above on the smoothability of the stratification of it may now be assumed not only that the stratification is smooth but, furthermore, that for each stratum Y, the bundle map $TM \to \gamma^c_{n,k}|H$ restricts, on $T\mathring{Y}$ to a map of vector bundles $T\mathring{Y} \to n^r_T$, where $h(Y) \subset X^r_T$. (In other words, the vector-bundle reduction of $T\mathring{Y}$ from which the smoothing arises comes from this bundle map.)

We are finally in a position to construct the piecewise-differentiable immersion $M^n \to R^{n+k}$ whose existence is the conclusion of 7.5.

We proceed, as is customary, by an induction based on the dimension of strata. For Y of dimension 0, we shall immerse, in fact embed, an open neighborhood of $M(Y)$ in R^{n+k} as follows: send Y (a single point) to the origin of R^{n+k}. The image $h(Y)$ is a point L in $\mathcal{L}_T$ and we may identify $M(Y)$ with the standard unit n-disc D_{U_L} in a standard way, and by extending radially slightly, we embed a slightly larger neighborhood $M(Y)$ in R^{n+k}. Obviously, the Gauss map on $M(Y)$ (which is a smoothly LS-stratified manifold) is h (up to a deformation of h via a stratum-preserving isotopy of M^n).

Now assume that there is an immersion $f_0: B \to R^{n+k}$ of a neighborhood B of $\bigcup_{\dim Z<i} M(Z) = M^{(i-1)}$ so that

(1) $g(f_0) = h$ on B.

Let Y be an i-stratum. We shall extend the immersion to f on a neighborhood of $B' \cup M(Y)$ so that $h \sim g(f)$ rel B' where $B' \subset B$ is a smaller neighborhood of $M^{(i-1)}$. First, we note that if $h(Y) \subset X^r_T$, we may deform h on $Y-B'$ so that $h(Y-B') \subset \bar{\mathcal{L}}^r_T$, $h(M(Y)-B') \subset N^r_T$. Note that $Y \cap (B-B')$ is smoothly immersed via f_0 in R^{n+k} so that for $y \in Y \cap (B-B')$ there is an $(n+k-i)$-dimensional affine subspace of U_y of R^{n+k} which is the normal space to $f_0(Y \cap (B-B'))$ at $f_0 y$. By definition, this coincides with $U_{g(f_0)(y)}$ where $g(f_0)(y) \in \bar{\mathcal{L}}^r_T \subset \mathcal{L}_T$ is to be thought of as a formal link.

Obviously, the translate to the origin of U_y (via $x \mapsto x-f_0 y$) is the subspace $U_{g(f_0)(y)}$. Now consider a small tubular neighborhood of $f_0(Y \cap B-B'))$, viewed as $\bigcup_y D_{\varepsilon,y}$ where $D_{\varepsilon,y}$ is the disc of radius ε in U_y. We may identify $\mathring{D}_{\varepsilon,y}$ with $\mathring{D}_{U_{g(f_0)(y)}}$ in a standard way. On the other hand, we have a map $g(f_0)\colon M(Y) \to M_T^r$ whose image lies in $\bigcup_y \mathring{D}_{U_{g(f_0)(y)}} \subset M_T^r$. The composite of $g(f_0)$ with the map given by the union of identifications $\mathring{D}_{\varepsilon,y} \longleftrightarrow \mathring{D}_{U_{g(f_0)(y)}}$ immerses $M(Y) \cap (B-B')$ in R^{n+k}. In fact we claim that, up to a slight regular homotopy of f_0 rel $M^{(i-1)}$, which does not alter $g(f_0)$, this immersion of $M(Y) \cap B$ coincides with f_0. Now let B'' be a neighborhood of $M^{(i-1)}$ intermediate between B and B' (i.e. $\bar{B}' \subset B''$, $\bar{B}'' \subset B$). We may deform h, in a stratum-preserving way, rel $\bar{B}''$, so that $h(Y-B') \subset \bar{\mathcal{Z}}_T^r \subset X_T^r$. Note that on $Y \cap (B-B')$ the composite $Y \cap (B-B') \xrightarrow{h} \bar{\mathcal{Z}}_T^r \xrightarrow{\theta_T} G_{i,n+k-i}$ is the standard Gauss map of a smooth immersion. Since $\theta_T \circ h\colon Y-B' \to G_{i,n-i}$ is covered by a vector-bundle map $T(Y-B') \to \alpha$, where α denotes the canonical i-vector bundle over $G_{i,n+k-i}$, it follows from the Hirsch immersion theorem [H] that there is a smooth immersion $\phi\colon Y-B' \to R^{n+k}$, coinciding with f_0 on $Y \cap (\bar{B}''-B')$ whose Gauss map is homotopic, rel $Y \cap (\bar{B}''-B')$ to $\theta_T \circ h$. In fact, if we let $G = \theta_T(\bar{\mathcal{Z}}_T^r)$ which is, by assumption, an open set, then the Gauss map ψ of ϕ has image in G and is homotopic to $\theta_T \circ h$ in G (rel the same subspace). To see this, we recall the extension of Hirsch's immersion result via Gromov-Phillips theory [P]. Now by the defining properties of geometric subspaces of $\mathcal{G}^c_{n,k}$, it follows that the deformation of $\theta_{T\circ h}$ to ψ may be covered by a deformation of $h|(Y-B')$ to some map h, $(Y-B') \to \bar{\mathcal{Z}}_T^r$ again rel $Y \cap \bar{B}''$. This deformation may be extended, rel $\bar{B}''$ to a deformation of h on M which preserves the stratification of M and the transversality condition. It is now clear that the immersion on Y may be extended to $M(Y)$ via the same argument as before. I.e. for each $y \in Y-B''$, we have a normal $n+k-i$ plane U_y, and the identification of $\mathring{D}_{U_{h(y)}}$ with $\mathring{D}_{\epsilon,y}$, when composed with h on $M(Y)$ produces an immersion, coinciding with f_0 on $M(Y) \cap B''$. Obviously, this immersion be "radially" extended to a small neighborhood $\hat{M}(Y)$ of $M(Y)$ (so that $B'' \cup \hat{M}(Y)$ is open). Let f be this extension; clearly $g(f) = h$ on $B'' \cup \hat{M}(Y)$ (h meaning the original h as it has been modified to this point by deformations). Finally, we note that this procedure may be carried out simultaneously for all i-dimensional Y, to produce a piecewise-differentiable immersion f on a neighborhood C of $M^{(i)}$ whose Gauss map $g(f)\colon C \to H$ coincides with h (i.e. with a deformed version of the original h). The inductive step is complete, and with it, the proof of the main theorem 7.5.

8. Some applications to smoothing theory

In this section we shall examine an interesting criterion for the smoothing of a PL manifold M^n, whose proof is based on the constructions of the previous sections.

First, consider the set of equivalence classes of linearly ordered combinatorial triangulations of the (i-1)-sphere (equivalence means order preserving simplicial equivalence). Given such an object represented by Σ^{i-1}, we have derived objects, or faces, one for each simplex σ of Σ^{i-1}. That is, we define Σ_σ as $\ell k(\sigma,\Sigma)$ with the inherited ordering and note that this passes over to equivalence classes. We see that $\dim \Sigma_\sigma = \dim \Sigma - \dim \sigma$.

We may now construct a CW complex in the by-now-familiar fashion: For each equivalence class $[\Sigma^{i-1}]$ we have one i-cell to be thought of as $c\Sigma^{i-1}$. (There is one zero cell for the -1-dimensional null set.) Identifications are made, as usual, by the face homeomorphisms $h_\sigma : c\Sigma_\sigma \to \Sigma \subset c\Sigma$ which identified $c\Sigma_\sigma$ with the dual cell $\sigma^* \subset \Sigma$ in the standard way. After making these identifications we obtain a CW complex which we label A^{ord}.

Note that as in §1, a locally-ordered triangulated combinatorial manifold M^n admits a canonical Gauss map $g: M^n \to A^{ord}$. (More precisely, g is defined only on $M^n_0 = \bigcup \sigma^*$, $\sigma \not\subset \partial M^n$ being a simplex of M^n.) The definition of g is the usual one; briefly, we send the dual cell σ^* to the cell of A^{ord} which is viewed, abstractly, as $c\,\ell k(\sigma,M^n)$ (i.e., the cell which corresponds to the equivalence class of $\ell k(\sigma,M^n)$ as an ordered triangulated sphere).

Now consider ordered triangulated (i-1)-spheres together with linear (on simplices) embeddings $\rho: c\Sigma,* \to R^i,0$ such that $\rho(v) \in S^{i-1}$ for v a vertex of Σ. We shall call two such pairs (Σ_1,ρ_1), (Σ_2,ρ_2) <u>equivalent</u> if and only if there is an order-preserving simplicial isomorphism $\phi: \Sigma_1 \to \Sigma_2$ such that

$\rho_1 = \alpha\cdot(\rho_2\circ\phi)$ where α is an element of the orthogonal group $O(i)$.

As before, face operations are defined. Given $[(\Sigma^{i-1},\rho)]$, and the simplex σ of Σ^{i-1} we obtain $(\Sigma_\sigma,\rho_\sigma)$ by taking Σ_σ to be the sphere $\ell k(\sigma,\Sigma)$ as an ordered complex, and by specifying ρ_σ as follows: Let U be the affine space of dimension $i\text{-}\dim\sigma-1$ orthogonal to $\rho(c\sigma)\subset R^i$ and passing through the barycenter b of $\rho(c\sigma)$. For v a vertex of $\ell k(\sigma)$, let v' be the intersection of U with the ray $\overline{0v}$ and v'' the radial projection of v' on the unit sphere of U centered at b. Extend convexly the assignment $v\mapsto v''$, $x\mapsto b$ to a linear embedding ρ' of $c\,\ell k(\sigma,\Sigma)$ in U. We identify U isometrically with an $(i\text{-}\dim\sigma-1)$-dimensional vector subspace of R^i by the usual translation $-b$ and thus, by composing with ρ', we obtain an embedding $\rho_\sigma: c\Sigma_\sigma\subset R^j$, $j = i\text{-}\dim\sigma-1$. Of course, there is an ambiguity here which arises from the indeterminacy of the identification of $U\text{-}b$ with R^j. But since this isometry is well-defined, up to the action of $O(j)$, it is clear that the equivalence class $[(\Sigma_\sigma,\rho_\sigma)]$ is well-defined. Moreover, this class clearly depends only on $[(\Sigma,\rho)]$ and not on (Σ,ρ) itself.

Thus, in complete analogy to our previous construction of A^{ord}, we obtain a C-W complex A^{Br} which has one i-cell (to be thought of as $c\Sigma^{i-1}$) for each equivalence class of $(i-1)$-dimensional ordered Brouwer spheres. Naturally, there is a forgetful map $A^{Br}\to A^{ord}$ induced by the assignment $(\Sigma^{i-1},\rho)\to\Sigma^{i-1}$, which is obviously consistent with equivalence relations. We note in passing that this forgetful map is not onto, inasmuch as there are combinatorially triangulated spheres which are not Brouwer spheres in that the respective cones do not embed linearly in Euclidean space with codimension 0.

We may retopologize A^{Br} to obtain a new space A^{cBr} in analogy to the retopologization of $\mathcal{G}_{n,k}$, which produced $\mathcal{G}^c_{n,k}$. That is, given two classes $[(\Sigma_1,\rho_1)]$ $[(\Sigma_2,\rho_2)]$ of $(i-1)$-dimensional ordered Brouwer spheres, consider them ε-close if there is an order preserving isomorphism $h\colon \Sigma_1 \to \Sigma_2$ such that ρ_1 is ε-close to $\rho_2 \circ h$ in the usual sense. Call two points $x, y \in A^{Br}$ ε-close in the new metric, if they have pre-images $\bar{x}$, $\bar{y}$ in $c\Sigma_1$, $c\Sigma_2$ respectively for two ordered Brouwer spheres (Σ_1,ρ_1), (Σ_2,ρ_2), if (Σ_1,ρ_1) is ε close to (Σ_2,ρ_2) via h, as above, and if $\rho_1\bar{x}$ is within ε of $\rho_2\bar{y}$ in Euclidean space. With this new metric A^{Br} acquires a smaller topology A^{cBr} and we obtain a forgetful map $A^{Br} \to A^{cBr}$ which is the identity pointwise. Again, we note that A^{Br} is naturally the geometric realization of a simplicial set having one j-simplex $\sigma([(\Sigma,\rho)],\tau)$ for each class of ordered Brouwer spheres $[(\Sigma,\rho)]$ and each j-simplex τ of $c\Sigma$. A^{cBr} is obtained from A^{Br} by topologizing the set of j-simplices, all j, in the obvious way and taking the geometric realization of the corresponding simplicial space.

Now note further that the forgetful map $A^{Br} \to A^{ord}$ factors as $A^{Br} \to A^{cBr} \xrightarrow{f} A^{ord}$, where the first map is the pointwise-identity. Our smoothing result concerns lifting of the diagram

$$\begin{array}{ccc} & & A^{cBr} \\ & & \downarrow f \\ M^n & \xrightarrow{g} & A^{ord} \end{array}$$

where M^n is a PL manifold with a locally-ordered combinatorial triangulation and g the natural map to A^{ord}.

<u>8.1 Theorem</u>. In the diagram above, there exists a homotopy lifting $h\colon M^n \to A^{cBr}$ if and only if there is a smoothing of the PL structure on M^n. In fact, a homotopy class of such liftings h determines a homotopy-class of vector bundle reductions of the stable tangent bundle of M^n, and thus a concordance class of smoothings of

M^n. Furthermore, each smoothing of M^n arises from at least one such lifting.

<u>8.2 Corollary</u>. With M^n as above, M^n is smoothable if there is a lifting $k: M^n \to A^{Br}$ in the diagram

$$\begin{array}{ccc} & & A^{Br} \\ & & \downarrow f \\ M^n & \xrightarrow{g} & A^{ord} \end{array}$$

We start our proof of 8.1 by investigating a certain geometric subspace of $\mathcal{G}^c_{n,k}$. Consider a formal link L of dimension $(n,k;j)$. Say that L is <u>straight</u> if $c\Sigma_L \subset U_L$ actually lies in some j-dimensional sub-vector space of U_L. Clearly, straightness of links is preserved under the face operations, thus $\mathcal{G}^{st}_{n,k} = \bigcup_{L \text{ straight}} e_L$ is a subcomplex of $\mathcal{G}_{n,k}$. Equally clearly, this is a geometric subcomplex.

When we retopologize $\mathcal{G}_{n,k}$ to obtain $\mathcal{G}^c_{n,k}$, we find that the image of $\mathcal{G}^{st}_{n,k}$ is a geometric subspace $\mathcal{G}^{cst}_{n,k}$ of $\mathcal{G}^c_{n,k}$. Moreover, we find that the restriction $\gamma_{n,k}|\mathcal{G}^{cst}_{n,k}$, has a canonical structure as an n-vector bundle. That is, if $x \in e_L \subset \mathcal{G}^{cst}_{n,k}$, then the fiber of $\gamma^c_{n,k}$ over x may be identified, as a vector space, with $U_L \oplus W_L$ where W_L is the unique j-plane containing $c\Sigma_L$ $(\dim L = j)$.

Note that $^{cst}_{n,k}$ contains the canonical image of $G_{n,k}$ in $\mathcal{G}^c_{n,k}$ (as the retopologized 0-skeleton of $\mathcal{G}_{n,k}$) and that the canonical n-plane bundle over $G_{n,k}$ is the restriction of the natural vector bundle structure on $\gamma^{cst}_{n,k} = \gamma_{n,k}|\mathcal{G}^{cst}_{n,k}$. The following is obvious:

<u>8.3 Lemma</u>. The PL manifold M^n is smoothable if and only if there is a piecewise-smooth immersion $f: M^n \to R^{n+k}$ such that

$g(f)(M^n) \subset \mathcal{G}^{cst}_{n,k}$.

To obtain a proof of 8.1, it is useful to revise some of our principal constructions to accommodate orderings. First of all, an <u>ordered</u> formal link (of dimension $(n,k;j)$) is a formal link $L = (U_L, \Sigma_L)$ together with a linear ordering on the vertices of Σ. Henceforth, the specific ordering shall be implicitly subsumed under the notation Σ_L, and we shall speak of the ordered link $L = (U_L, \Sigma_L)$. As before, we obtain, for each simplex σ of Σ_L, the derived ordered link L_σ. The usual construction obtains for us a CW complex $\mathcal{G}^{ord}_{n,k}$ with one j-cell e_{L_0} for each ordered link L of dimension j. In passing, we note that $\mathcal{G}^{ord}_{n,k}$ is the natural target of the Gauss map for linear immersions of triangulated, locally ordered manifolds. Moreover, there is a natural forgetful map $\mathcal{G}^{ord}_{n,k} \to A^{ord}$.

Now we may retopologize $\mathcal{G}^{ord}_{n,k}$, just as was done with $\mathcal{G}_{n,k}$, to obtain $\mathcal{G}^{cord}_{n,k}$. Again, we note briefly how $\mathcal{G}^{cord}_{n,k}$ fits into the scheme of things as the natural target of a Gauss map. We consider smoothly LS-stratified manifolds M^n such that, for each stratum X of M^n, the triangulated sphere $\ell k(X)$ is ordered and, moreover, the orderings are consistent with incidence relations. That is, if $Y < X$, the natural inclusion $\ell k(X) \subset \ell k(Y)$ is order preserving. It is understood that ordering of $\ell k(X)$ labels the local incidences of X, i.e. if $M(X)$, $X_0 = M(X) \cap X$ are in in §7, then we demand that for each component Y_r of $Y \cap M(X)$, with $\dim Y = \dim X + 1$, $X < Y$, there is a specific vertex $v(Y_r)$ of $\ell k(X)$. Thus, viewing $M(X)$ as a $c\ell k(X)$ bundle over X_0, we see that it is trivialized in a specific way. Denote such manifolds as smoothly-LOS stratified manifolds (LOS = linkwise ordered simplicial). Given a piecewise-smooth immersion of the smoothly LOS-stratified manifold M^n in R^{n+k} (satisfying α, β, γ of §7.3 as usual) we obtain a Gauss map

$M^n \to \mathcal{G}^{cord}_{n,k}$.

The analogues of theorems 4.2 and 7.5 may easily be obtained by simple modifications of the respective definitions and proofs, but this is not our major concern.

Consider now the analog of $\mathcal{G}^{st}_{n,k}$, that is, the geometric subcomplex $\bigcup_{L \text{ straight}} e_L$ where the union is taken over the set of ordered links whose underlying unordered links are straight in the sense just described. Denote this complex by $\mathcal{G}^{os}_{n,k}$. Its image in $\mathcal{G}^{cord}_{n,k}$ is denoted $\mathcal{G}^{cos}_{n,k}$ and, as in the case of $\mathcal{G}^{cst}_{n,k}$, we see that the natural PL bundle $\gamma^{cord}_{n,k}$ over $\mathcal{G}^{cord}_{n,k}$ has a restriction $\gamma^{cos}_{n,k} = \gamma^{cord}_{n,k} | \mathcal{G}^{cos}_{n,k}$ which admits a natural n-dimensional vector bundle structure. (The existence of natural PL n-plane bundles over $\mathcal{G}^{ord}_{n,k}$, $\mathcal{G}^{cord}_{n,k}$ etc. is seen by obvious extensions of the construction of $\gamma_{n,k}$, $\gamma^{c}_{n,k}$ in §2 and §7 respectively.)

Now we note that the forgetful map $\mathcal{G}^{ord}_{n,k} \to A^{ord}$ prolongs to $\mathcal{G}^{cord}_{n,k} \to A^{ord}$. Moreover, we see that there is a natural map $\mathcal{G}^{os}_{n,k} \to A^{Br}$. We see this on the cell level by assigning to the j-link L, the class of ordered Brouwer spheres $[(\Sigma,\rho)]$ where Σ, is abstractly, Σ_L as an ordered simplicial complex, and where $c\Sigma$ is embedded in R^j by taking the vertices of Σ_L, plus the origin, and convexly extending to obtain a linear embedding of $c\Sigma \cong c\Sigma_L$ in U_L, which embedding, by definition of straightness, factors as $c\Sigma \subset W \subset U_L$ where W is a j-dimensional subspace. This gives us an embedding $\rho: c\Sigma \to R^j$ when we pick any linear isometry of W with R^j, and the class $[(\Sigma,\rho)]$ is unaffected by this choice. Hence we may send the cell e_L of $\mathcal{G}^{os}_{n,k}$ to the cell of A^{Br} corresponding to $[(\Sigma,\rho)]$. This assignment essentially defines the map $\mathcal{G}^{os}_{n,k} \to A^{Br}$. (It is a simple exercise to show that this assignment of cells is consistent with inclusion of faces.)

We make the further observation that the map above respects the retopolization of $\mathcal{G}^{os}_{n,k}$ as $\mathcal{G}^{cos}_{n,k}$ and of A^{Br} as A^{cBr}, thus producing the natural diagram

$$\begin{array}{ccc} \mathcal{G}^{os}_{n,k} & \to & \mathcal{G}^{cos}_{n,k} \\ \downarrow & & \downarrow \\ A^{Br} & \to & A^{cBr} \end{array}$$

It is appropriate now to point out that there are natural stabilizing maps in n and k, $\mathcal{G}^{X}_{n,k} \to \mathcal{G}^{X}_{n+1,k}$, $\mathcal{G}^{X}_{n,k} \to \mathcal{G}^{X}_{n,k+1}$ where X is any of the superscripts that have been introduced to this point. The definition follows the example of $\mathcal{G}_{n,k}$ and $\mathcal{G}^{c}_{n,k}$ without essential change. We use $\mathcal{G}^{X}$ to denote the direct limit $\varinjlim_{n,k} \mathcal{G}^{X}_{n,k}$.

We note that the natural maps $\mathcal{G}^{ord}_{n,k} \to A^{ord}$, $\mathcal{G}^{cord}_{n,k} \to A^{ord}$, $\mathcal{G}^{os}_{n,k} \to A^{Br}$, $\mathcal{G}^{cos}_{n,k} \to A^{cBr}$ commute with the stabilizing maps in n and k. Thus we have, in fact, natural maps $\mathcal{G}^{ord} \to A^{ord}$, $\mathcal{G}^{cord} \to A^{ord}$, $\mathcal{G}^{os} \to A^{Br}$, $\mathcal{G}^{cos} \to A^{cBr}$. To prove 8.1, we shall study the diagram

$$\begin{array}{ccc} \mathcal{G}^{cos} & \to & A^{cBr} \\ \downarrow & & \downarrow \\ \mathcal{G}^{cord} & \to & A^{ord} \end{array}$$

The key idea is to look at the action of the infinite orthogonal group $O = \lim_{m} O(m)$ on $\mathcal{G}^{cos}$ and $\mathcal{G}^{cord}$ respectively. The action arises in the following way: given n and k, we consider an ordered link L of dimension $(n,k;j)$. An element α of $O(n+k)$ transforms L in the following way (as in §5): If $L = (U_L, \Sigma_L)$, then αL is specified by $(\alpha U_L, \alpha\Sigma_L)$ where, in the first instance αU_L denotes the image of U_L via the natural $O(n+k)$ action on $G_{j+k,n-j}$ and in the second $\alpha\Sigma_L$ means merely the image of Σ_L under the action of α acting as an isometry on R^{n+k}. We assume of course that the vertices of $\alpha\Sigma_L$ are ordered corresponding to the ordering of Σ_L. This action on the set of formal links defines, for each $\alpha \in O(n+k)$, a cellular homeomorphism $\alpha: \mathcal{G}^{ord}_{n,k} \to \mathcal{G}^{ord}_{n,k}$, but, as we observed in §5, we do not have an $O(n+k)$-action on $\mathcal{G}^{ord}_{n,k}$, in the sense of a continuous map $O(n+k) \times \mathcal{G}^{ord}_{n,k} \to \mathcal{G}^{ord}_{n,k}$. Rather, we have an action of $O(n+k)$ with the discrete topology. Our first observation, however, is that when $\mathcal{G}^{ord}_{n,k}$ is retopologized to obtain $\mathcal{G}^{cord}_{n,k}$, the $O(n+k)$ action does become continuous in the usual sense. Verification of this fact is straightforward and details are left to the reader.

Furthermore, this $O(n+k)$ action preserves straightness of links, thus $\mathcal{G}^{os}_{n,k}$ and $\mathcal{G}^{cos}_{n,k}$ are invariant under the action. Next, we see that under the natural inclusion $O(n+k) \to O(n+k+1)$ the action of $O(n+k)$ commutes with the stabilizing maps $\mathcal{G}^{X}_{n,k} \to \mathcal{G}^{X}_{n+1,k}$, $\mathcal{G}^{X}_{n,k} \to \mathcal{G}^{X}_{n,k+1}$, X = cos, cord.

If we look at the maps $\mathcal{G}^{cord}_{n,k} \to A^{ord}$, $\mathcal{G}^{cos}_{n,k} \to A^{cBr}$, we find that these maps are $O(n+k)$-maps if we make A^{ord}, A^{Br} $O(n+k)$-spaces via the trivial action. Thus we have maps

$$\begin{array}{ccc} \mathcal{G}^{cos}_{n,k}/O(n+k) & \to & A^{cBr} \\ \downarrow & & \downarrow \\ \mathcal{G}^{cord}_{n,k}/O(n+k) & \to & A^{ord} \end{array}$$

Passing to the limit in n and k, we have the diagram

$$\begin{array}{ccc} \mathcal{G}^{cos}/O & \overset{\mu}{\dashrightarrow} & A^{cBr} \\ \downarrow & & \downarrow \\ \mathcal{G}^{cord}/O & \overset{\nu}{\dashrightarrow} & A^{ord} \end{array}$$

<u>8.4 Lemma.</u> μ and ν are homotopy equivalences.

Proof: To deal with μ first we claim that it is, in fact, a homeomorphism. Clearly, it is onto, since, given a linear embedding ρ of $c\Sigma^{i-1},* \to R^i,0$ taking vertices to S^{i-1}, were Σ^{i-1} is ordered, we may take Σ' to be the corresponding geodesic triangulation of S^{i-1}, thus obtaining the formal ordered link (obviously straight) $L = (R^i,\Sigma')$ of dimension $(i,0;i)$. Obviously $\mu(e_L)$ is the cell of A^{Br}, (and thus the subspace of A^{Br}) corresponding to the equivalence class of the ordered Brouwer sphere (Σ,ρ).

Now suppose x,y are points in $\mathcal{G}^{\cos}/0$ with $\mu(x) = \mu(y)$. Let $\bar{x},\bar{y}$ denote preimages of x and y in $\mathcal{G}^{\cos}$ and thus in $\mathcal{G}^{\cos}_{n,k}$ for suitably large n and k. Then $\bar{x},\bar{y}$ are, respectively, in the interior cells e_L,e_K where L and K are of the same dimension j. Thus we may think of $\bar{x}$ and $\bar{y}$ as points of $c\Sigma_L$, $c\Sigma_K$ respectively. Now denote by Σ_L', Σ_K', respectively, the isomorphic simplicial complexes linearly embedded in R^{n+k} by taking, e.g. a typical simplex of Σ_L' to be the convex hull of the vertices of a typical simplex of Σ_L. There are standard homeomorphisms $\Sigma_L \overset{a_L}{\longleftrightarrow} \Sigma_L'$ and $\Sigma_K \overset{a_K}{\longleftrightarrow} \Sigma_K'$ which extend to standard homeomorphisms of the respective cones. Let $\bar{\bar{x}}$ and $\bar{\bar{y}}$ be the corresponding points to $\bar{x}$ and $\bar{y}$ in $c\Sigma_L'$, $c\Sigma_K'$ respectively. Recall that the condition that L and K be straight links means, in the first place that $c\Sigma_L$ (resp. $c\Sigma_K$), and therefore $c\Sigma_L'$ (resp. $c\Sigma_K'$) lie in necessarily unique j-planes W_L, W_K respectively. Recall also that the image of points of $c\Sigma_L'$ (resp. $c\Sigma_K'$) in A^{cBr} under the composition $c\Sigma_L' \overset{a_L}{\longleftrightarrow} c\Sigma_L \to e_L \subset \mathcal{G}^{\cos}_{n,k} \to \mathcal{G}^{\cos}/0 \overset{\mu}{\longrightarrow} A^{cBr}$ is determined by identifying W_L (resp. W_K) with R^j by an arbitrary isometry ϕ_L thus obtaining the ordered brouwer sphere Σ_L, $\phi_L|c\Sigma_L$ and thus obtaining an appropriate cell of A^{Br} which maps into A^{cBr} under retopologization. Now if $\bar{\bar{x}}$ and $\bar{\bar{y}}$ go to the same point in A^{Br}, it must necessarily be because there is an element $\psi \in O(j)$ such that ψ induces an ordered simplicial isomorphism $\phi_L(\Sigma_L') \to \phi_K(\Sigma_K')$ with $\psi(\phi_L\bar{\bar{x}}) = \phi_K\bar{\bar{y}}$. But then $\phi_K^{-1}\psi\phi_L$ is an isometry $W_L \to W_K$ which induces the corresponding isomorphism $\Sigma_L' \to \Sigma_K'$ with $\phi_K^{-1}\psi\phi_L\bar{\bar{x}} = \bar{\bar{y}}$. Now extend $\phi_K^{-1}\psi\phi_L$ to an element α of $O(n+k)$ with $\alpha U_L = U_K$. Obviously $\alpha L = K$. Moreover, since $\alpha\bar{\bar{x}} = \bar{\bar{y}}$, it follows, by the consistency of the identifications $c\Sigma_L \overset{ca_L}{\longleftrightarrow} c\Sigma_L'$, $c\Sigma_K \overset{ca_K}{\longleftrightarrow} c\Sigma_K'$ with orthogonal transformation that $\alpha\bar{x} = \bar{y}$. Thus $\bar{x}$ and $\bar{y}$ are identified by the action of $O(n+k)$ on $\mathcal{G}_{n,k}^{\cos}$ and hence $x = y$.

The remaining point is that μ is an open map. This is left to the reader as an exercise.

The argument for ν is somewhat more delicate. First, let T denote an isomorphism class of ordered triangulated $(j-1)$-spheres, and let $\mathcal{K}_T$ be the union of all ordered formal links L of dimension j with $\Sigma_L \in T$. I.e. if $\mathcal{L}_T(n,k)$ is the set of all formal links L of dimension $(n,k;j)$ with $\Sigma_L \in T$, then $\mathcal{K}_T = \bigcup_{n,k}\mathcal{L}_T(n,k)$, it being understood that a link of dimension $(n,j;j)$ is to be identified with its suspensions which are $(n+1,k;j)$ and $(n,k+1;j)$-dimensional links. Now of course $\mathcal{L}_T(n,k)$ has a certain topology, namely that which was used in obtaining the topology of $\mathcal{G}_{n,k}^{\mathrm{cord}}$. Thus $\mathcal{K}_T$ acquires the weak topology of union. Now the action of $O(n+k)$ on links of dimension $(n,k;j)$ preserves $\mathcal{L}_T(n,k)$ as an invariant subspace. In the limit, we have an action of the infinite orthogonal group O on $\mathcal{K}_T$. We wish, first of all to characterize the homotopy type of $\mathcal{K}_T/O$.

<u>8.5 Lemma</u>. $\mathcal{K}_T/O$ is (weakly) contractible.

Proof: Let n and k both be large compared to the number of vertices of a representative of T. We wish to study $\mathcal{L}_T(n,k)/O(n+k)$. Let R^{j+k} denote the standard $j+k$-space in the standard R^{n+k} and let $\mathcal{J}_T(n,k)$ be the space of those $L \in \mathcal{L}_T(n,k)$

such that $U_L = R^{j+k}$.

<u>8.6 Sublemma.</u> Any $L \in \mathcal{L}_T(n,k)$ is in the same $O(n+k)$ orbit as some element of $\mathcal{J}_T(n,k)$. Moreover any two elements of $\mathcal{J}_T(n,k)$ are in the same $O(n+k)$ orbit if and only if they are in the same $O(j+k)$ orbit. Therefore $\mathcal{L}_T(n,k)/O(n+k) = \mathcal{J}_T(n,k)/O(j+k)$.

Proof: First of all, for any $L = (U_L, \Sigma_L) \in \mathcal{L}_T(n,k)$ we may find an $\alpha \in O(n+k)$ so that $\alpha U_L = R^{j+k}$, thus $\alpha L \in \mathcal{J}_T(n,k)$. Next, if $L_1, L_2 \in \mathcal{J}_T(n,k)$ and $L_2 = \alpha L_1$ for some $\alpha \in O(n+k)$, then R^{j+k} is an invariant subspace for α and thus $\alpha = \alpha_1 \oplus \alpha_2 \in O(j+k) + O(n-j)$ where the latter summand acts on $(R^{j+k})'$. But then $\alpha' = \alpha_1 \oplus I_{n-j}$ has the same effect on L_1 as α did, thus L_2 differs from L_1 by an action of $O(j+k)$. The remainder of the sublemma follows immediately.

Note that throughout the proof of 8.5 we have kept j fixed and, moreover, $\mathcal{J}_T(n,k)$ is independent of n and depends only on k and j. Thus we modify notation and write $\mathcal{J}_T(k)$ for $\mathcal{J}_T(n,k)$. In the light of 8.6 we need only show that $\lim_{k\to\infty} {}_T(k)/O(j+k)$ is weakly contractible.

Let $\mathcal{J}_T = \lim \mathcal{J}_T(k)$, and let $L \in \mathcal{J}_T$. Define the positive integer span L to be the dimension of the vector space spanned by the vertices of Σ_L, taken as vectors of Euclidean space. Span L is independent of the choice of the $\mathcal{J}_T(k)$ in which to regard L as belonging, and is obviously invariant under the action of $O(j+k)$.

<u>8.7 Sublemma.</u> For $L \in \mathcal{J}_T(k)$, the isotropy group of L under the action of $O(j+k)$ on $\mathcal{J}_T(k)$ is $O(j+k\text{-span } L)$; thus the orbit of L under this action is of type $O(j+k)/O(j+k\text{-span } L)$ and is thus $(j+k\text{-span } L)$-connected. Thus, in the limit $\mathcal{J}_T$, the orbit of L is contractible.

Proof: Let Z_L be the subspace spanned by the unit vectors which are the vertices of Σ_L. If α is in the isotropy group of L, it must leave Z_L pointwise fixed since no vertex of Σ_L may be moved. On the other hand, so long as this restriction is met, any action is possible on the invariant subspace $Z_L^{\perp}$ without compromising membership in the isotropy group. I.e. the isotropy group consists of all elements of $O(j+k)$ leaving Z_L pointwise fixed, i.e. $O(j+k-\dim Z_L) = O(j+k-\text{span } L)$. The remainder of 9.7 follows immediately.

<u>8.8 Sublemma</u>. $\mathcal{J}_T$ is (weakly) contractible.

Proof: $\mathcal{J}_T$ is, essentially the space of embeddings of Σ in S^∞ as an admissibly-triangulated subsphere, where Σ is a fixed representative of T. The contractibility of this space was demonstrated in Lemma 7.2

To summarize, $\mathcal{K}_T/O = \lim \mathcal{L}_T(n,k)/O(n+k) = \lim \mathcal{J}_T(k)/O(j+k) = \mathcal{J}_T/O$. In this last quotient space, the numerator $\mathcal{J}_T$ is weakly contractible, while the orbit of every point is weakly contractible. It follows that $\mathcal{J}_T/O$ is contractible, at least weakly, thus, our lemma 8.5 follows.

Returning to the proof that u is a homotopy equivalence, we investigate the geometry of $\mathcal{G}^{\text{cord}}$ and its orbit space $\mathcal{G}^{\text{cord}}/O$. Just as in the previous chapter, $\mathcal{L}_T(n,k)$ may be identified with the union in $\mathcal{G}^{\text{cord}}_{n,k}$ of cone points of $e_L = \text{image } c\Sigma_L$, the union being taken over all $L \in \mathcal{L}_T(n,k)$. Also as before, $\mathcal{L}_T(n,k)$ has a neighborhood in $\mathcal{G}^{\text{cord}}_{n,k}$, denoted N_T where N_T is the image of a (trivial) j-disc bundle P_T over $\mathcal{L}_T$. (Here $j = \dim T$; the fiber of P_T over $L \in \mathcal{L}_T$ is, of course, identified with Σ_L. Let $Q^{(j)}_{(n,k)} = \bigcup_{\dim T \leq j-1} N_T \subset \mathcal{G}^{\text{cord}}_{n,k}$. Then we see that $Q^{(j+1)}_{(n,k)}$ is $Q^{(j)}_{(n,k)}$ plus some spaces of the form $\mathcal{L}_T(n,k) \times D^{j+1}$ attached by maps $\mathcal{L}_T(n,k) \times S^j \to Q^{(j)}$. Passing to the limit, $\mathcal{G}^{\text{cord}}$, we easily see that

$\mathcal{K}_T = \bigcup_{n,k} \mathcal{L}_T(n,k)$ is embedded and has a neighborhood $\mathcal{N}_T$ which is the image of a trivial j-disc bundle $\mathcal{P}_T$ over $\mathcal{K}_T$. Again, if we let $Q^{(j)} = \varinjlim_{n,k} Q^{(j)}(n,k) = \bigcup_{\dim T<j-1} \mathcal{N}_T \subset \mathcal{G}^{cord}$, we see that the same characterization of $Q^{(j+1)}$ holds, viz.

$Q^{(j+1)} = Q^{(j)} \cup \bigcup_{\dim T=j} \mathcal{K}_T \times D^{j+1}$ where the products are attached to $Q^{(j)}$ by maps on $\mathcal{K}_T \times S^j$.

Now, when we pass to $\mathcal{G}^{cord}/0$ it is interesting to note that the same sort of picture emerges: The image of $\mathcal{N}_T$, denoted by $\hat{\mathcal{N}}_T$ is a j-disc bundle over $\mathcal{K}_T/0$, mod identifications on the bounding sphere bundle. That is, $\mathcal{P}_T$ is naturally induced from a bundle $\hat{\mathcal{P}}_T$ over $\mathcal{K}_T/0$ where $\hat{\mathcal{P}}_T = \mathcal{P}_T/0$. (This claim essentially reduces to the fact that the action of 0 cannot identify two points within the interior of the same e_L, while it must identify e_L with e_K in some fashion if it identifies L with K.) $\hat{\mathcal{P}}_T$ maps naturally into $\mathcal{G}^{cord}/0$ via a map which is a homeomorphism off the bounding sphere bundle. Thus we have once more a filtration: namely $\hat{Q}^{(j)} \subset \mathcal{G}^{cord}/0$ is defined as $\bigcup_{\dim T<j-1} \hat{\mathcal{N}}_T$, and once more we have

$\hat{Q}^{(j+1)} = \hat{Q}^{(j)} \cup \bigcup_{\dim T=j} (\ _T/0)\times D^{j+1}$ via attaching maps on the bounding $(\mathcal{K}_T/0)\times S^j$.

We are finally able to finish the proof of 8.4 by showing μ to be a homotopy equivalence. Recall that A^{ord} has one j-call for each equivalence class of ordered triangulated (j-1)-spheres. Let T be such a class: then think of the call corresponding to T as cT where, by slight abuse of notation we confuse T with a representative. In the space $\mathcal{K}_T/0\times D^j = \hat{\mathcal{P}}_T$ mentioned in the preceeding paragraph, each tiber is, in fact, identified with cT in a specific way, i.e. we have a canonical map $m_T: \mathcal{K}_T/0\times D^j \to cT$. This map preserves the decomposition of $\mathcal{G}^{cord}/0$ and A^{ord} respectively into the spaces $\mathcal{K}_T/O\times D^j$ and cT. That is, the identifications on $\mathcal{K}_T/O\times D^j$ are consistent, under m_T, with the identifications on T. Thus the m_T taken together yield a map $\mathcal{G}^{cord}/O \to A^{ord}$, which is in fact, μ. Now assume inductively that $\mu|\hat{Q}^{(j)}$ is a homotopy equivalence onto the j-skeleton $A^{(j)}$ of A^{ord}. Attach $\hat{\mathcal{P}}_T$ to $\hat{Q}^{(j)}$ for some j-dimensional T, while attaching cT to $A^{(j)}$. Clearly the effect of adding $\hat{\mathcal{P}}_T$ to $\hat{Q}^{(j)}$ is merely to add a $(j+1)$-cell to $\hat{Q}^{(j)}$, up to homotopy, since $\hat{\mathcal{P}}_T$ is a (weakly) contractible space crossed with D^{j+1}. Now m_T clearly maps this homotopy cell by a degree one map to the cell cT. Clearly then, $\hat{Q}^{(j)} \cup \hat{\mathcal{P}}_T$ is homotopically the same as $A^{(j)} \cup cT$ since the cells were added, in either case, by equivalent attaching maps. The same argument still holds when we attach all the $\hat{\mathcal{P}}_T$ simultaneously, to obtain $\hat{Q}^{(j+1)}$. Taking the limit $\lim \hat{Q}^{(j)} = \mathcal{G}^{cord}/O$, we see that μ must be a homotopy equivalence, as desired, and 8.4 is thus proved.

Continuing with the proof of 8.1, we examine the diagram

$$\begin{array}{ccccccc} F^{cos} & \to & \mathcal{G}^{cos} & \to & \mathcal{G}^{cos}/O & \sim & A^{cBr} \\ \downarrow & & \downarrow & & \downarrow & & \downarrow \\ F^{cord} & \to & \mathcal{G}^{cord} & \to & \mathcal{G}^{cord}/O & \sim & A^{ord} \end{array}$$

where F^{cos} and F^{cord} are the fibers of the respective quotient maps $\mathcal{G}^X \to \mathcal{G}^X/O$, $X = cos, cord$.

<u>8.9 Lemma</u>. $F^{cos} \to F^{cord}$ is a homotopy equivalence.

Proof: We first look at the action of $O(n+k)$ on $\mathcal{G}^{cos}_{n,k}$, and in particular, at the orbit types which occur under this action.

First of all, it is clear that if $x \in \operatorname{int} e_L \subset \mathcal{G}^{cos}_{n+k}$, then the isotropy group I_x of x is exactly the isotropy group I_L of L

itself under the action of $O(n+k)$ on the space of straight ordered links. However, it is easily seen that for a j-dimensional straight link L, I_L consists of those orthogonal transformations α which, first of all leave U_L invariant and which, moreover, leave W_L, the j-plane containing $c\Sigma_L$, pointwise fixed. I_L is thus conjugate to $O(k)\times O(n-j)$, so the same is true of I_x. Now let L,K be straight ordered links of dimension j_1,j_2 respectively with $L = K_\sigma$ for some simplex σ of Σ_K. Then, clearly $I_K \subset I_L$; we wish to characterize this inclusion, and we see, as the reader may easily check, that up to conjugacy it looks like $O(k)\times O(n-j_2) \to O(k)\times O(n-j_1)$ where the map is given by the identity on the $O(k)$-factor and the standard inclusion $O(n-j_2) \subset O(n-j_1)$ on the other. Thus, if $y \in \text{int } e_K$, $x \in \text{int } e_L$ then $I_y \subset I_x$ may be similarly characterized. Hence the connectivity of the inclusion $I_y \subset I_x$ is a function solely of n, k, j_1 and j_2, and, assuming j_1,j_2 fixed, this connectivity goes to ∞ with n and k.

Extending this analysis to the action on $\mathcal{G}^{cord}_{n,k}$, once more we see that, for $x \in \text{int } e_L$, $I_x = I_L$. Let $j = \dim L$ $s = \text{span } L$. Then we see that I_L is conjugate to $O(k+j-s)\times O(n-j)$. Again, if $L = K_\sigma$ where the respective dimensions are j_1,j_2 and spans k_1,k_2, then $I_K \subset I_L$ is conjugate to the product of the standard inclusions $O(k+j_2-s_2) \subset O(k+j_1-s_1)$ and $O(n-j_2) \subset O(n-j_1)$ (note in this regard that $s_2-j_2 < s_1-j_1$). Thus, if $x \in \text{int } e_L$ and $y \in \text{int } e_K$, then the inclusion $I_y \subset I_x$ may be similarly characterized. As before, we see that the connectivity of $I_y \subset I_x$ depends only on a few parameters, namely n, k, j_1, j_2, s_1, s_2, and, keeping the last four (which depend only on the geometry of Σ_L and Σ_K) constant we see that this connectivity goes to ∞ with n and k.

Passing, then, to the limit actions of O on $\mathcal{G}^{cos}$ and $\mathcal{G}^{cord}$, we see that all orbit types are homotopy equivalent, i.e. $I_y \subset I_x$

is a homotopy equivalence when $x \in \text{int}\ e_L$ and $y \in e_L$. Thus the homotopy type of the orbit type O/I_x is constant. By taking x to be the unique point of the 0-cell e_L for L a 0-dimensional link we see that $O/I_x = \lim_{n,k} O(n+k)/O(k) \times O(n) = BO$, the usual classifying space for stable vector bundles. It is thus possible to deduce that in the fibration $F^{cos} \to \mathcal{G}^{cos} \to \mathcal{G}^{cos}/O$ the abstract homotopy-theoretic fiber F^{cos} is identifiable with $BO \simeq O/I_x$ for all x. The same holds true for the fibration $F^{cord} \to \mathcal{G}^{cord} \to \mathcal{G}^{cord}/O$ and, of course, $F^{cos} \to F^{cord}$ is a homotopy equivalence. This completes 8.9.

It follows immediately that, given a diagram

$$\begin{array}{ccc} & & \mathcal{G}^{cos} \\ & \overset{\gamma}{\nearrow} & \downarrow \beta \\ X & \xrightarrow{\alpha} & \mathcal{G}^{cord} \end{array}$$

where X is a C-W complex, homotopy classes of liftings γ are in 1-1 correspondence with homotopy classes of liftings γ' in the diagram

$$\begin{array}{ccccc} & & \mathcal{G}^{cos}/O & \sim & A^{cBr} \\ & \overset{\gamma'}{\nearrow} & \downarrow \beta' & & \downarrow \beta' \\ X & \xrightarrow{\alpha'} & \mathcal{G}^{cord}/O & \sim & A^{ord} \end{array}$$

where α' is the composition of α with projection and β' is the pushdown of the O-equivariant map β.

Now let M^n be an ordered triangulated manifold. Immerse M^n in R^{n+k} by a piecewise smooth (e.g. linear on simplices) map thus obtaining a Gauss map $\phi: M^n \to \mathcal{G}^{cord}_{n,k} \subset \mathcal{G}^{cord}$. It is easily seen that, up to homotopy within $\mathcal{G}^{cord}$, ϕ does not depend on the triangulation (within the PL equivalence class of M^n) nor on the ordering nor the codimension k.

Now if $\phi': M^n \to A^{ord}$ lifts to A^{cBr}, we shall have a lifting of ϕ to $\mathcal{A}^{cos}$. But since $\gamma^c|\mathcal{A}^{cos}$ has a vector bundle structure and ϕ induces the stable tangent bundle of M^n from γ^{cord} it follows that M^n is smoothable; in fact the given lifting of ϕ' induces a particular lifting of ϕ, hence a specific stable vector bundle reduction of TM thus a particular smoothing of M^n. Note, however, that $\phi': M^n \to A^{ord}$ is easily identified up to homotopy with the standard map $M^n \to A^{ord}$ defined at the beginning of this section, which depends merely on the local structure of M^n as a locally-ordered triangulated manifold. To see that any smoothing of M^n comes from such a lifting, we endow such a smoothing with a smooth triangulation, locally order that triangulation, and immerse M^n smoothly in R^{n+k}. The Gauss map ϕ is then seen to have its image in $\mathcal{A}^{cos}$, and thus ϕ' strictly factors in a unique way through A^{cBr}. Thus we have a standard lift of ϕ' to A^{cBr} which clearly induces the given smoothness structure.

This completes the proof of 8.1.

<u>8.10 Remark</u>. The reader should note that nowhere did we claim that the natural bundle structure γ^{cord} over $\mathcal{A}^{cord}$ is induced from a bundle on A^{ord}. We do not claim that the PL bundle data prolong to A^{ord}, but merely that the smoothing problem does!

<u>8.11 Remark</u>. Note that we do not claim a 1-1 correspondence between liftings of

$$\begin{array}{ccc} & & A^{cBr} \\ & & \downarrow f \\ M^n & \xrightarrow{g} & A^{ord} \end{array}$$

and smoothings of M^n; we merely assert that the existence of one is a necessary and sufficient condition for existence of the other. In fact, liftings of g classify structures on M^n somewhat richer than mere smoothings. Consider, for the given ordered triangulated M^n, structures on M^n given by LOS-stratifications of M^n, concordant to the triangulation together with smoothings of M^n so that each stratum is a smooth submanifold. Call two such structures equivalent if and only if the two smoothings are concordant while the two LOS-stratifications are concordant through a concordance whose strata are smooth submanifolds of the concordance of smoothings. Then homotopy classes of liftings in the diagram

$$\begin{array}{ccc} & & A^{cBr} \\ & & \downarrow f \\ M^n & \xrightarrow{g} & A^{ord} \end{array}$$

are in 1-1 correspondence with equivalence classes of such structures.

<u>8.12 Remark</u>. Consider the fiber of $A^{cBr} \xrightarrow{f} A^{ord}$, i.e. the fiber P of ${}^{cos} \to {}^{cord}$. We claim that, on the level of homotopy groups at least, this fiber has PL/O as a summand. To see this, consider the diagram

$$\begin{array}{ccc} P & \longrightarrow & PL/O \\ \downarrow & & \downarrow \\ \mathcal{G}^{cos} & \xrightarrow{\gamma_1} & BO \\ \downarrow & & \downarrow \\ \mathcal{G}^{cord} & \xrightarrow{\gamma} & BPL \end{array}$$

where γ classifies the natural PL bundle γ^{cord} over $\mathcal{G}^{cord}$ and γ_1 the natural vector-bundle structure of $\gamma^{cord}|\mathcal{G}^{cos}$. We claim that there are inverse maps

$$\begin{array}{ccc} BO & \xrightarrow{s} & \mathcal{G}^{cos} \\ \downarrow & & \downarrow \\ BPL & \xrightarrow{t} & \mathcal{G}^{cord} \end{array}$$

splitting and $_1$. Of course there is a natural map $BO \to \mathcal{G}^{cos}$ which results from identifying $G_{n,k}$ pointwise, with the 0-skeleton of $\mathcal{G}^{ord}_{n,k}$ and hence, topologically, with the image of that skeleton in $\mathcal{G}^{cord}$. However, to define s and extend it to t, we must use a somewhat different map. Think of BO as approximated by a high-dimensional smooth manifold V, with the universal stable vector bundle being approximated by TV. Now think of BPL as approximated by a PL manifold W whose tangent bundle approximates the universal stable PL bundle. We may assume that V is a submanifold of of W. Triangulate W combinatorially so that V is a subcomplex with a smooth triangulation. Immerse W piecewise-smoothly in Euclidean space, so that V is smoothly immersed. Then the Gauss map, stabilized, gives us a map $W,V \to \mathcal{G}^{cord}, \mathcal{G}^{cos}$ which we may take as an approximation of s and t. It follows that $\pi_*(PL/O)$ is a summand of $\pi_* P$.

The reader may find it instructive to compare our result with the approach of Cairns and Whitehead [C_1, C_2, C_3; Whd] to smoothing theory for combinatorial manifolds. This approach, it will be recalled, involves the idea of a transverse field of k planes on a manifold M^n embedded in R^{n+k}. The existence of such a transverse field is shown by Cairns, with gaps repaired by Whitehead, to guarantee the existence of a smooth structure on M^n. The problem of finding such a field, when M^n is linearly embedded with respect to some combinatorial triangulation, is then analyzed to a certain extent. The nub of this analysis is that if σ is a j-simplex of M^n, and $st(\sigma,M^n)$ is in "general position" (i.e., in our terminology, the formal link $L(\sigma,M^n)$ has maximal span equal to the number of vertices of $\ell k(\sigma,M^n)$), then the space of k-planes transverse to M^n at a point s of σ is homeomorphic to the space of linear embeddings $c\ell k(\sigma,M^n),* \quad R^{n-j},*$, divided out by the action of the general linear group $GL(n-j;R)$. Of course, this latter space up to homotopy, is the same as the space of linear embeddings $c\ell k(\sigma, M^n, *) \to R^{n-j}$ which take vertices of $\ell k(\sigma, M^n)$ to S^{n-j-1}, divided out by the action of $O(n-j)$. Call this space $Br(\ell k(\sigma, M^n))$.

In our approach to smoothing, the same space obviously plays a role, that is given a cell e of A^{ord}, corresponding to the ordered sphere Σ^{j-1}, its inverse image in A^{cBr} is a trivial j-disc bundle over the space $Br(\Sigma^{j-1})$. In fact, it may easily be seen that a transverse field to the triangulated $M^n \subset R^{n+k}$ yields, if M^n be locally ordered, a section h in the diagram

$$\begin{array}{ccc} & & A^{cB_r} \\ & \overset{h}{\nearrow} & \downarrow f \\ M^n & \xrightarrow{g} & A^{ord} \end{array}$$

so that $h(\sigma^*)$ $f^{-1}g(\sigma^*)$ for each dual cell σ^* of M^n. Our theorem, 8.1 of course, merely asks for a homotopy section, and thus, not for a specific transverse field.

Our result would seem to invite renewed interest in the topology of the spaces $Br(\Sigma)$. Recently, D. Henderson has analyzed completely the case for Σ of dimension 2 [He].

Our result also invites comparison with the first part of Telemann's paper [T] on the "differential geometry" of PL manifolds. In this secion, Telemann characterizes geometrically the PL/O-bundle over a PL manifold M^n, sections of which are in 1-1 correspondence with smoothings of M^n. His approach seems interestingly analogous to the one adopted here.

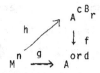

9. Equivariant Piecewise Differentiable Immersions

As we have seen in §5, if π is a finite group acting orthogonally on R^{n+k}, there is an induced action of π on $\mathcal{G}_{n,k}$. It is immediately evident that under the retopologization of the underlying point set which converts $\mathcal{G}_{n,k}$ to $\mathcal{G}^c_{n,k}$, the action of π remains continuous. Thus, it is natural to study PL n-manifolds M^n supporting locally smooth π-actions with a view towards developing some results on necessary and sufficient conditions for equivariant piecewise-differentiable immersions in R^{n+k} which respect additional geometric constraints. These results will be in large measure analogous to those of §5.

By way of background, consider an immersion of the sort contemplated in §7. That is, M^n is a smoothly LS-stratified manifold, and $f: M^n \to R^{n+k}$ is an immersion satisfying the conditions α, β, γ of §7. Now let us suppose that π acts on R^{n+k} orthogonally and on M^n (locally-smoothly in the sense of Bredon [Br]) so that the action is a group of self-equivalences from the point of view of LS-stratified manifolds. This means that if $p \in \pi$ then $p: M^n \to M^n$ preserves the stratification and is a diffeomorphism of each stratum to its image stratum. The conclusion we obviously want is thatthe Gauss map $g(f): M^n \to \mathcal{G}^c_{n,k}$ is equivariant. There is however, the slightly subtle point that $g(f)$ is not, strictly speaking, well-defined. It is dependent, be it remembered, upon choice of the decomposition $\{M(X)\}$ of M^n (X ranging over the strata), and the stratum preserving map $u: M^n \to M^n$. However, it may easily be ascertained that the constructions of $\{M(X)\}$ and u can be done so as to respect the action of π (i.e., u is equivariant and $pM(X) = M(pX)$ for p π. This, in turn, renders $g(f)$ equivariant.

Moreover, a final nicety is to note that two choices of equivariant $\{M(X)\}$ and u differ by equivariant ambiant isotopy, thus the resulting Gauss maps for f are not only equivariantly homotopic, but, in fact, made so by this isotopy.

On the bundle level, it will be apparent that the action of π on $\mathcal{G}^c_{n,k}$ further extends to a continuous action on $\gamma^c_{n,k}$. Again, since the total space of $\gamma^c_{n,k}$ coincides on the set level with that of $\gamma_{n,k}$, the action of π on the total space is immediately specified. Continuity may be checked routinely. Thus $\gamma^c_{n,k}$ acquires a π-bundle structure over the π-space $\mathcal{G}^c_{n,k}$.

<u>9.1 Proposition</u>. If $f: M^n \to R^{n,k}$ is an equivariant immersion (with respect to the action of π, the bundle map $TM^k \to \gamma^c_{n,k}$ covering the Gauss map $g(f) \to \mathcal{G}^c_{n,k}$ is equivariant.

We omit the proof.

The problem which will occupy us for the remainder of this section is that of deriving the analog of Theorem 5.2. Let H be a geometric subspace of $\mathcal{G}^c_{n,k}$; which is invariant under the action of π. Let M^n be an open PL manifold with a locally smooth action by π satisfying the Bierstone condition.

<u>9.2 Theorem</u>. If $h: M^n \to H$ is an equivariant map covered by a π-bundle map then there is an equivariant LS stratification of M^n so that, with respect to this stratification, there exists a piecewise-differentiable and equivariant immersion $f: M^n \to R^{n+k}$ (satisfying (α), (β) and (γ) of §7) such that $g(f): M^n \to \mathcal{G}^c_{n,k}$ has its image in H and $g(f)$ is equivariantly homotopic to h in H.

We shall give a proof which, in its essentials, follows the outline of 7.5 with appropriate modifications to deal with the action of π as needed. There is, however, one point in the proof of 7.5 which, however trivial in the absence of a group action, is clearly not trivial in the present case. This has to do with transversality arguments. We are in fact referring to that section of the proof of

7.5 which, by way of preliminary, induces an LS-stratified structure on M^n. it will be recalled that this argument, in turn, was essentially drawn from Lemma 7.4. Upon examination, one sees that the heart of the matter is that the map $h: M \to H$ can be made transverse, simultaneously to all the $\bar{\mathcal{L}}_T$'s via a small deformation of h (keeping M in H). Our problem is to recapitulate this argument in an equivariant context. Thus the equivariant transversality problem must be analyzed away before the rest of the proof may proceed.

9.1 Equivariant transversality

The key result we shall need involves finding conditions sufficient to allow equivariant transversality arguments to go through. In our case, we shall be dealing with maps equivariant with respect to the action of π, (the domain being a PL manifold). The range, in turn, contains a π-invariant subspace with an invarient neighborhood, and the aim will be to deform the map equivariantly so as to become transverse to the subspace.

Let B be a π-space and $p: E \to B$ a π-bundle over B (i.e., a PL disc bundle). There is a natural bundle over E, viz. $\xi = p^*(p)$, and, since p is equivarient, this is also a π-bundle. Now let M be a PL manifold on which π acts and $f: M \to E$ a π-equivariant map. Suppose further that M satisfies the Bierstone condition.

<u>9.3 Lemma</u>. Suppose $f_0: M \to E$ is covered by a π-bundle map $F: TM \to \xi \oplus \alpha$, where α is a π-bundle over E; then f is equivariantly deformable to $f_1: M \to E$ with g transverse to B. Moreover, there will be a π-bundle map $T(g^{-1}B) \to \alpha|B$.

Proof: Assume, without loss of generality that B is a π-manifold (of some dimension g). E.g., one might replace the original B by its equivariant regular neighborhood for some

π-embedding in the sphere of an orthogonal representation of π.

Also, regard a map $f: M \to E$ as a section $\bar{f}$ of $\bar{E} = M \times E \to M$.

Since TM, in our case, is the sum $TM = f_0^*(\alpha \oplus \xi) = f_0^*\alpha \oplus f_0^*\xi$, we may think of $f_0^*\xi$ as a sub-PL-bundle θ (equivariant under π) of TM, and we denote its fiber at x by θ_x.

Now, given any $\bar{f}$ arising from $f: M \to E$, consider a PL map (not a bundle map) ϕ: $TM \to T\bar{E}$ covering $\bar{f}$. Let $\bar{B} = M \times B \subset M \times E = \bar{E}$. We say that ϕ is nice if for every fiber θ_x, $\phi|\theta_x$ is transverse to $\bar{B}$. This makes sense, since, for $x \in M$, $T\bar{E}_{\bar{f}x}$ may be identified with a small neighborhood of $\bar{f}_x$, and with such a neighborhood small enough, the condition on $\phi|\theta_x$ is independent of how this is done.

Given $\bar{f}$, there is a natural map $D\bar{f}$: $TM \to T\bar{E}$, essentially the inclusion of $T\bar{f}(M)$ into $T\bar{E}|\bar{f}(M)$. Clearly, if $D\bar{f}$ is nice, $\bar{f}$ is transverse to $\bar{B}$, i.e. f is transverse to B.

Note that niceness of maps ϕ is an open condition. In the case at hand, consider $\bar{f}_0$; we have a π-bundle map $TM \to \alpha \oplus TE$. Let $\bar{\alpha}$, $\overline{TE}$ denote pullback to $\bar{E}$, and note that there is a projection $\bar{\alpha} \oplus \overline{TE} \to \overline{TE}$. Thus there is a composite $\phi_0: TM \to \bar{\alpha} \oplus \overline{TE} \to \overline{TE}$ covering $\bar{f}_0$. Note that ϕ_0 is nice, in our sense. Thus, by the extension of Bierstones equivariant version [Bi] of Gromov-Phillips theory, $\bar{f}_0$ is equivariantly deformable to a section $\bar{f}_1$ such that $D\bar{f}_1$ is nice and the corresponding map $f_1: M \to E$ has the required transversality property. Moreover, on $f_1^{-1}B = V$ the tangent bundle is given by the "complement" of θ_x at each point x. More formally if we consider $TM|V$ and $\theta|V$ as a subspace, it is clear that the normal bundle of θ in TM restricts on V to the tangent bundle TV of V. But, by this description, it is seen that TV is identified with $f_1^*\alpha|V = f_0^*\alpha|V$.

9.4 Remark. There is, as well, a relative version of 9.3. We replace the hypotheses of 9.3 by the assumption that $f_o : M \to E$ is already transverse to B on the codimension-0 submanifold M_o, and that M, M_o satisfies the relative Bierstone condition. In this case we obtain an equivariant deformation, rel M_o, to the desired transverse regular map f_1.

The usefulness of 9.3 and 9.4 appears in the first stage of the proof of 9.2. It will be recalled that the analogous first stage of the proof of 7.5 involved showing that the map $h: M^n \to H$, (H a geometric subspace of $\mathscr{G}^c_{n,k}$) induced an LS-stratification of M. The proof came about by appeal to transversality, completely unproblematical in the case where no group action is involved. We wish to make the analogous argument in the presence of the action of π on M, $\mathscr{G}^c_{n,k}$ and its subspace H.

We merely sketch the proof. The inductive aspects merely follow the pattern set by the analogous step in 7.5 (which rests in turn on the argument of 7.4). In the following, we understand that M may be replaced, at need, by a multiply equivariantly punctured version, since the argument outlined will, in the long run, produce the desired LS-stratified structure on such a punctured manifold. We then find a π-homeomorphic copy of the original M inside which is, a fortiori LS-stratified in the desired manner. Assume, therefore, that for all triangulation-classes T of formal links of dimension $r < j$ we have (for $\bar{\mathscr{L}}_T = \mathscr{L}_T \cap H$, $\bar{N}_T = N_T \cap H$:

(1) $h^{-1}\bar{\mathscr{L}}_T$ is a codimension r submanifold.

(2) $h^{-1}(\bar{N}_T)$ is a bundle neighborhood with

$$\begin{array}{ccc} h^{-1}(\bar{N}_T) & \to & \bar{N}_T \\ \downarrow & & \downarrow \\ h^{-1}\bar{\mathscr{L}}_T & \to & \bar{\mathscr{L}}_T \end{array}$$

a bundle map (in fact, of π-bundles). Not let T be of dimension j. We let $M_T = h^{-1}(\bar{N}_T)$. Omitting arguments, we claim that M_T may be taken to be an invariant codimension-0 submanifold of M. We may assume h is transverse to $\bar{\mathcal{Z}}_T$ near $\bigcup_{\dim U<j} \bar{N}_U$. We note the bundle map $TM_T \to \gamma^c_{n,k}|N_T$ is, in effect a map to the sum $\alpha \oplus \xi$, where α is the canonical Γ-vector bundle over $\mathcal{L}_T$ restricted to $\mathcal{Z}_T = \mathcal{L}_T \cap H$ and where ξ is the pullback under projection $p\colon \bar{N}_T \to \mathcal{Z}_T$ of this bundle itself. We puncture M_T equivariantly if necessary, to make it satisfy the Bierstone condition, relative to a region near $\bigcup_{\dim U<j} N_U$ and away from $\mathcal{Z}_T$. The Lemma 9.3 and its relative form 9.4 then apply and we make h transverse to $\bar{\mathcal{Z}}_T$ by an equivariant deformation inside $\bar{N}_T$. The deformation is constant outside a co-dimension-0 submanifold of M_T. The last step is to further equivariantly deform h so that a bundle neighborhood of $h^{-1}\bar{\mathcal{Z}}_T$ in M_T "expands" to fill all of $\bar{N}_T$, with the remainder of M_T being pushed outside of $\bar{N}_T$ without, however, intersecting $\bigcup_{\dim U<j} N_U$.

The remainder of the proof of 9.2 may now be carried through closely mimicking the proof of the non-equivariant version. It will be recalled that there are two further principal steps: Smoothing the stratification and actually constructing the desired immersion. The first is carried out as in §7 using Lashof-Rothenberg equivariant smoothing theory [La-Ro] in place of Cairns-Hirsch since each invariant stratum of M with link T produced by the argument just above has a tangent bundle induced from some π-vector bundle (i.e. from the canonical π-vector bundle over $\mathcal{L}_T$). The second may be done by using the Bierstone result in place of the Hirsch immersion theorem. Note that to make these arguments run, we may have to further puncture M equivariantly (in order to make each equivariant stratum satisfy the Bierstone condition rel a neighborhood of its boundary.) So we thereby immerse a multiply equivariantly-punctured M. But, as has been pointed out before, a π-homeomorphic copy of the original M may be found inside this.

Finally, we note that the main result 9.2 of this section has been stated for <u>open</u> π-manifolds (rather than manifolds with boundary) satisfying the Bierstone condition. A further technical condition will allow the extension of the result to manifolds with boundary (essentially by allowing the transversality arguments to go through on the boundary). We will not pursue this elaboration at this point.

10. Piecewise Differentiable Immersions into Riemannian Manifolds

In §6 we considered how to construct a complex $\mathcal{G}_{n,k}(W)$ for a triangulated $(n+k)$-manifold W in order to study PL immersions into W, by way of extending results on PL immersions into R^{n+k}. In this section we perform an analogous extension for piecewise smooth immersions. It is natural that the role played by the triangulation of W in §6 is now assumed by a smooth Riemannian structure on W.

We first show how a space $\mathcal{G}^c_{n,k}(W)$ may be built which plays a role analogous to $\mathcal{G}_{n,k}(W)$ in the prior study.

With W^{n+k}, as indicated, a Riemannian manifold, consider, for any point $w \in W$, the tangent space T_wW with its metric. An isometry of T_wW with R^{n+k} allows us to consider formal links of dimension $(n,k;j)$, $0 < j < n$, associated to T_wW. That is, a formal link L of dimension j in T_wW is a pair (U,Σ) where U is a $j+k$-plane in T_wW, and Σ an admissibly-triangulated $(j-1)$-sphere in the unit sphere S_U. As we have seen in §7, the set of all such links of dimension j has a natural topology so that the union of all links of all dimensions becomes a simplicial space. We now topologize the union of all j-dimensional links over all w W. This is done in the obvious way. First note that the Riemannian metric on W naturally metrizes TW. Moreover, for a fixed dimension j, the classical Grassmannian bundle $G_{j+k,n-j}(W) = \bigcup_w G_{j+k,n-j}T_wW$ also acquires a natural metric. Let $L = (U_L, \Sigma_L)$, $L' = (U'_{L'}, \Sigma'_{L'})$ be j-dimensional links associated respectively to the points $w, w' \in W$. They are ε-close if there is a simplicial isomorphism $\phi: \Sigma_L \to \Sigma'_L$ such that

(1) for all vertices v of Σ, ϕv is ε-close to v in TW.

(2) U is ε-close to U' in $G_{j+k,n-j}(W)$.

As usual we form the set $\bigcup c\Sigma_L$, over all links associated to all points $w \in W$, identifying $c\Sigma_{L,\sigma}$ with its image under $h(L,\sigma)$ for a simplex $\sigma \subset \Sigma_L$. The image of $c\Sigma_L$ is, as usual, denoted e_L. We topologize so that two points x,y are close if they are images, under the identification map of points $\bar{x}$, $\bar{y}$ in $c\Sigma_L$, $c\Sigma_K$ respectively where L is ε-close to K and $\bar{x}$, $\bar{y}$ are ε-close in TW. Call the resulting space $\mathcal{G}^c_{n,k}(W)$. Note the abvious fact that $\mathcal{G}^c_{n,k}(W) = \bigcup_{w \in W} \mathcal{G}^c_{n,k}(w)$ where $\mathcal{G}^c_{n,k}(w)$ is a copy of $\mathcal{G}^c_{n,k}$ arising under the isometry $T_wW \cong R^{n+k}$. In fact $\mathcal{G}^c_{n,k}(W)$ is a fiber bundle over W with fiber $\mathcal{G}^c_{n,k}$.

We next construct the canonical PL n-disc bundle $\gamma^c_{n,k}(W)$ over $\mathcal{G}^c_{n,k}(W)$. It is obvious that we want to have

$\gamma^c_{n,k}(W) = \bigcup_x \gamma^c_{n,k}(x)$ where $\gamma^c_{n,k}(x)$ is a copy of $\gamma^c_{n,k}$ over $\mathcal{G}^c_{n,k}(x)$. However, for our purposes, it is most useful to use the following picture of the PL bundle structure of $\gamma^c_{n,k}(W)$.

Again, let T denote an isomorphism class of $(j-1)$-dimensional triangulated spheres. We now use $\mathcal{L}_T(W)$ to denote the topological space whose points are j-dimensional formal links L (at all points of W) such that $\Sigma_L \in T$. We let $\mathcal{L}_T(x)$ be those links of $\mathcal{L}_T(W)$ based at x. Thus $\mathcal{L}_T(x)$ is a copy of $\mathcal{L}_T \subset \mathcal{G}^c_{n,k}$ and $\mathcal{L}_T(W)$ is thus naturally embedded in $\mathcal{G}^c_{n,k}(W)$. Recall also the space N_T, the image mod some identification on the sphere bundle boundary of the natural PL j-disc bundle P_T over $\mathcal{L}_T$. N_T, it will be recalled is naturally included in $\mathcal{G}^c_{n,k}$. Thus, if we let $N_T(w)$ denote the corresponding copy in $\mathcal{G}^c_{n,k}(w)$, we may define $N_T(W)$ as $\bigcup_{w \in W} N_T(w)$.

It will be seen that $N_T(W)$ is naturally the image in $\mathcal{G}^c_{n,k}(W)$ of a PL j-disc bundle over $\mathcal{L}_T(W)$. That is, if we take $\bigcup_{L \in \mathcal{L}_T(W)} c\Sigma_L = P_T(W)$, with the appropriate topology we have $\mathcal{L}_T$

embedded as the union of all cone points, making $\mathcal{L}_T$ the 0-section of a PL j-disc bundle. $N_T(W)$ is the image of $P_T(W)$ under the quotient map which takes $c\Sigma_L \to e_L$ for all L.

Recall now the natural map $\mathcal{L}_T \to G_{n-j,k+j}$ of §7, defined by the map $L \mapsto U_L^{\perp}$, L a j-dimensional link. The natural generalization shows that there is a map $s_T: \mathcal{L}_T(W) \to G_{n-j,k+j}(W)$. In analogy to 7 we have

<u>10.1 Lemma</u>. The map $s_T: \mathcal{L}_T(W) \to G_{n-j,k+j}(W)$ is a fibration.

Note that under the map $P_T(W) \to N_T(W)$ two points may be identified only if they lie over the same point of $\mathcal{L}_T(W)$. Thus there is a projection $N_T(W) \to \mathcal{L}_T(W)$. Therefore we may define a bundle γ_T over $N_T(W)$ as the pullback of $\alpha \oplus \beta$ over $\mathcal{L}_T(W)$ where α is the PL bundle $P_T(W) \to \mathcal{L}_T(W)$ and β is the pullback, via $\mathcal{L}_T(W) \to G_{n-j,k+j}(W)$, of the canonical $(n-j)$-vector bundle over $G_{n-j,k+j}(W)$. γ_T is $\gamma^c_{n,k}|(N_T(W))$. It is straightforward to show that there is a natural way to glue γ_{T_1} to γ_{T_2} over $N_{T_1}(W) \cap N_{T_2}(W)$ for two triangulation classes T_1, T_2 (of different dimensions). This follows, since $\gamma_{T_i}|N_{T_i}(x) = \gamma^c_{n,k}(x)|N_{T_i}(x)$ so the gluing is well-defined by the manner in which the $\gamma_{T_i}|N_{T_i}(x)$ are glued together for each x.

Consider, therefore, a smoothly LS-stratified manifold M^n. Let $f: M^n \to W$ be an immersion piecewise smooth in the sense of §7. In line with our usual procedure, we define a Gauss map $g(f): M \to \mathcal{G}^c_{n,k}(W)$ covered by a bundle map $TM \to \gamma^c_{n,k}(W)$. Much as in §7, we consider the decomposition of M into subsets $M(X)$, one for each stratum X of M. As in §7 we proceed inductively. Assuming that $g(f)$ has been defined for $\bigcup M(X)$ over all strata X of dimension $< j$, we look at a typical j-dimensional stratum X, with the understanding that $M(X)$ is parameterized as a twisted product $X_o \times_\tau N_*^{\Sigma}$, $\Sigma = \mathcal{S}(X)$, or, more conveniently for our immediate purpose, as $\overset{\circ}{X}_o \times_\tau c\Sigma$ (the notation here is as in §7). Now, if $x \in \overset{\circ}{X}$, let $y = f(x) \in W$. The "differential" df may be understood as mapping $x \times c\Sigma$ to T_yW and is, in fact, a PL homeomorphism to its image. In fact, this image, together with the given simplicial structure of Σ, defines a formal link of dimension n-j in T_yW, denoted $L(x,f)$, which we think of as a point in $\mathcal{L}_T(y), T = [\Sigma]$. By this pointwise definition we obtain a continuous map $h: \overset{\circ}{X} \to \mathcal{L}_T(w)$. By composing with the map $U: X_o \to X$ we obtain $h \circ u$ on int X_o. Recall that $u|$ int X_o is covered by an obvious bundle map from the normal bundle of int X_o to the normal bundle of $\overset{\circ}{X}$. Since h is covered by a bundle map from the normal bundle of $\overset{\circ}{X}$ to the canonical bundle $P_T(W)$ over $\mathcal{L}_T(W)$ in a canonical way, we obtain a map int $X_o \times_\tau c\Sigma \to P_T(W) \to N_T(W)$, which defines the needed extension of $g(f)$ to $M(X)$.

Definition of the covering bundle map $TM \to \gamma_{n,k}^c(w)$ over $g(f)$ is also done by generalizing mildly the techniques of §7. Be it noted, in this regard, that on X_o this bundle map splits as a direct sum map $TX_o \oplus \nu X_o \to \eta_T \oplus P_T(W)$ where η_T denotes the canonical (j-vector) bundle induced by $s_T: \mathcal{}_T(W) \to G_{j,n+k-j}(W)$. $TX_o \to \eta_T$ is a vector-bundle map, with respect to the vector bundle structure on TX_o coming from the given smooth structure on the stratum X.

The next order of business is to generalize the notion of geometric subspace to the present context. We adapt it nearly verbatim from the corresponding definition in §7.

<u>10.2 Definition</u>. The subspace $H \subset \mathcal{G}_{n,k}^c(W)$ is said to be geometric under the following conditions

(1) If $H \cap \overset{\circ}{N}_T(W) \neq \emptyset$, then this intersection is an open disc bundle (with fiber $cT - T$) over $H \cap \mathcal{L}_T(W)$.

(2) $H \cap \mathcal{L}_T(W)$ is open in $\mathcal{L}_T$ and its image $s_T(H \cap \mathcal{L}_T(W))$ is

open in $G_{n-j,j+k}(W)$ Here, $j-1 = \dim T$ and s_T is the natural map $\mathcal{L}_T(W) \to G_{n-j,j+k}(W)$.

(3) $s_T | H \cap \mathcal{L}_T$ is a fibration.

As in §7, the most natural examples of geometric subspaces, in the light of 10.1, are of the form $\bigcup_{T\in\mathcal{V}} N_T(W)$ where $\mathcal{V}$ is some collection of equivalence classes of triangulations where, if $\Sigma \in T \in \mathcal{V}$ and v is a vertex of Σ, then $[\ell k(v,\Sigma)] \in \mathcal{V}$.

The main result of this section imitates 7.5.

<u>10.3 Theorem</u>. Let H be a geometric subspace of $\mathcal{G}^c_{n,k}(W)$. Let M^n be a PL manifold with no closed components. Suppose that the map $h\colon M^n \to H$ is covered by a bundle map $TM^n \to \gamma^c_{n,k}(W)|H$. Then M^n admits a stratification as an LS-stratified manifold such that, with respect to this stratification, there is a piecewise differentiable immersion $f\colon M^n \to W^{n+k}$ so that $g(f)\colon M^n \to \mathcal{G}^c_{n,k}(W)$ has its image in H and, moreover, $g(f)$ is homotopic to h in H.

There is, as well, a relative version analogous to 7.6 whose statement we omit.

The proof of 10.3 proceeds much in the manner of 7.5. We briefly indicate it here, paying particular attention to such modifications as may be needed.

If we adopt the notation $\bar{\mathcal{L}}_T = \mathcal{L}_T(W) \cap H$, then $\bar{N}_T = N_T(W)$ is a PL disc bundle over $\bar{\mathcal{L}}_T$. As in the proof of 7.5 we see that H can be "stratified" with one stratum for each component of $\bar{\mathcal{L}}_T$.

The first modification of the given map h lies in deforming it so as to be transverse to this stratification. We thereby induce a stratification on M^n wherein a typical stratum Y is a component of the inverse image of a stratum of H.

Smoothing of this stratification also follows the model set down in 7.5 with little modification.

The main idea, then, is to take the given smooth LS stratification of M^n and construct, with respect thereto, an immersion $f: M^n \to W^{n+k}$ which is piecewise-differentiable and whose Gauss map $g(f)$ is homotopic to h in H.

Note that, given h, there is an obvious map, $f_o : M^n \to W^{n+k}$ simply because $\mathcal{G}^c_{n,k}(W)$ is a fibre bundle over W, and thus f_o is given by $M \xrightarrow{h} H \subset \mathcal{G}^c_{n,k}(W) \xrightarrow{\text{proj}} W$. (In passing note that for f a piecewise-differentiable immersion, f is recovered from its Gauss map in precisely this way.) Clearly, in the case at hand, the immersion to be produced shall be homotopic to f_o.

As usual, we proceed inductively. Assume that f_o is an immersion on a neighborhood of $\bigcup_{\dim Z<i} M(Z) = M^{(i-1)}$, Z ranging over the strata of M. Assume, furthermore, that h restricted to this neighborhood is the Gauss map of the immersion. Now let Y be an i-stratum; clearly, our goal is to deform f_o, rel a slightly smaller neighborhood of $M^{(i-1)}$, so as to make it an immersion on a neighborhood of $M^{(i-1)} \cup M(Y)$. The Gauss map of the immersion will at the same time be homotopic to h rel the same smaller neighborhood. Moreover, it will be seen that the deformation of h near $M(Y)$ is stratum preserving, and thus extendable to a stratum-preserving deformation of h on all of M. If we then append the observation that the modification of h and of f_o can be carried out, in succession, for each i-stratum Y, we shall see that we have completed the inductive step, i.e., replaced $M^{(i-1)}$ by $M^{(i)}$ in the inductive hypothesis.

To proceed to the details of this inductive step, consider the i-dimensional stratum Y. Let N be the neighborhood of $M^{(i-1)}$ on which f_o is a piecewise-smooth immersion; let N' be a slightly smaller neighborhood with $\overline{N}'$ N. (We may take N, N' to be manifolds-with-boundary if we like.) Let Y_1, Y_2 be $Y \cap N'$, $Y \cap N$

respectively, and we may assume that $Y - Y_2 \subset Y_o$ is bounded away from Y_1. Now, over $Y_o \subset Y$ we have the PL-disc bundle $M(Y)$. If $Y_1' = Y_1 \cap Y_o$, $Y_2' = Y_2 \cap Y_o$, we may assume, without loss of generality that $M(Y)|Y_1'$ N, $M(Y)|Y_2' \subset N'$. (This may involve modifying the choice of N, N'). Thus we obtain an immersion of $M(Y)|Y_1'$.

We want now to make an assumption on this immersion. The statement involves recalling that, if Y has T as link, then $h: Y_o \to \bar{\mathcal{L}}_T$ and, moreover, the map of the tangent bundle $TM \to \gamma^c_{n,k}(W)$ is the "direct sum" of two maps, viz:

(1) $TY_o \to \xi$, where ξ is the vector bundle over $\bar{\mathcal{L}}_T$ coming from the canonical map $\mathcal{L}_T \to G_{i-ih+k+i}(W)$

(2) The map of PL disc-bundles (with discrete structure group) $M(Y) \to \overline{N}(T)$.

Thus, if $y \in Y_1'$, we have a fiber D_y of $M(Y)$ over y, and it is identified with an $(n-i)$-disc $D'_{h(y)}$ in the fiber disc of $\gamma^c_{n,k}(W)$ over $h(y)$. But, since, if $w = f_o y = \text{proj } h(y) \in W$, the fiber of $\gamma^c_{n,k}$ over $h(y)$ is viewed as a subspace of $T_W W$, we have essentially identified D_y itself with a sub-disc of the unit disc in $T_W W$.

Now we may clarify the assumption we wish to make: It is that for $y \in Y_1'$ sufficiently far from Y_2' (i.e., close to $Y_o - Y_1'$) the immersion $f_o|D_y$ coincides with the composition

$$D_y \cong D'_{h(y)} \subset T_W W \xrightarrow{\exp_\varepsilon} W$$

Here, it must be explained that $\varepsilon > 0$ and $\exp_\varepsilon$ denotes the exponential map for the Riemannian structure on W, reparameterized so that rays of length s in $T_W W$ go onto geodesics of length $\varepsilon \cdot s$.

We claim, without detailed proof, that we may modify h and f_o on N, rel N' so that this assumption is realized. The proof is by use of the standard tubular neighborhood theorem for the smoothly-embedded manifold Y_1'.

Proceeding, therefore, on the basis of this assumption, the remainder of the proof of 10.3 follows fairly rapidly. First of all, we note the map $Y \xrightarrow{h} \bar{\mathcal{Z}}_T \to G_{i,n+k-i}(W)$, which is covered by a vector bundle map from TY to the i-flag bundle of W. (The non-closed assumption on M is used, as in §7, to produce this bundle map along with the piecewise-smooth stratification of M). By the Hirsch immersion theorem [Hi], we may deform f_o on Y, rel a neighborhood of Y_2, to an immersion f_1. Moreover, this deformation is covered by a deformation of the bundle map from TY to the canonical i-plane bundle over $G_{i,h+k-i}(W)$, such that the terminal stage of the deformation coincides with the map arising from the differential of the smooth immersion. Additionally, we may, by virtue of Gromov-Phillips theory [P] insure, as well, that the entire deformation of the map $Y_o \to G_{i,n+k-i}(W)$ stays within the open set im $(\bar{\mathcal{Z}}_T)$. Since, by the definition of geometric subspace, $\bar{\mathcal{Z}}_T$ is fibered over its image in $G_{i,n+k-i}(W)$, we obtain a deformation of $h|Y_o \to \bar{\mathcal{Z}}_T$. Now cover this by a deformation of $h|M(Y) \to \bar{N}(T)$, and let h_1 denote the terminal stage of this deformation on both Y and $M(Y)$. For each Y in $Y_o - Y_2$, we have, as we have seen, an identification of D_y with $D'_{h_1(y)}$ and thus with a disc in $T_{f_1(y)}(W)$. Composing with $\exp_\varepsilon$ (ε varying continuously with y) we obtain an immersion on $M(Y)$ extending the given one on N^1; an easy argument suffices to extend this still further to an open neighborhood of $N' \cup M(Y)$. It is obvious from the construction that the Gauss map of this immersion is h_1 (i.e. h on N' and h_1 on $M(Y)$) and it is trivial that the deformation of h to h_1 (over an open neighborhood of $N' \cup M(Y)$) extends to a stratum preserving deformation of h on M.

Thus, the key inductive step is concluded and, by virtue of our earlier remarks, the main result 10.3 has been proved, at least in outline.

We conclude this section with some brief generalizations. First of all, it is natural to consider Riemannian manifolds W with a finite group T acting by isometries. In this case the obvious equivariant generalizations of the definitions and results of this section are easily accessialbe. We see immediately that given the isometric action of π on W the space $\mathcal{G}^c_{n,k}(W)$ becomes a π space and the canonical bundle $\gamma^c_{n,k}(W)$ a bundle with π-action. Furthermore, given a manifold M^n with π-action and an equivariant piecewise smooth immersion $M \to W$, we obviously obtain a Gauss map $M \to \mathcal{G}^c_{n,k}(W)$ which is π- equivariant. Now consider geometric subspaces H of $\mathcal{G}^c_{n,k}(W)$ invariant with respect to π. We then have an obvious generalization of previous results:

<u>10.4 Theorem</u>. Let M^n be a PL π-manifold satisfying the Bierstone condition and let H be an invariant geometric subspace of $\mathcal{G}^c_{n,k}(W)$. Suppose that there is a π-equivariant map $h: M^n \to H$ covered by a π-bundle map $TM^n \to \gamma^c_{n,k}|H$. Then M^n admits an equivariant LS-stratification such that, with respect to this stratification, there is an equivariant immersion $f: M^n \to W^{n+k}$ whose equivariant Gauss map $g(f)$ has image in H and is, in fact π-homotopic to h in H.

The proof combines techniques of this section with those of the previous section §9 where equivariant immersions into linear actions on Euclidean spaces were considered.

A concluding remark concerns a further generalization which will not even spell out in any detail. Suffice it to say that, for our target manifold W, we need not take a smooth Riemannian manifold but only a smoothly L·S stratified manifold whose strata are equipped with Riemannian metrics which make face inclusions isometries. It is an exercise in combining the constructions and techniques of this section with those of §6 to formulate a precise definition of $\mathcal{G}^{c}_{n,k}(W)$ in this case, to specify the kinds of immersions for which Gauss maps are defined, and to prove results analogous to 10.3, and, in the equivariant case, to 10.4.

APPENDIX: Glossary of Important Definition and Constructions

PAGE		REMARK
1.3	local ordered formula and local formula for characteristic classes.	
1.7-9	s-cell complex; s-block bundle; s-cell; $\mathcal{S}$-cell set.	"$\mathcal{S}$-cell sets" is a category equivalent to CW complexes but more useful in facillitating consideration of ordered combinatorial subdivision.
1.10	Q_n; $\lvert Q_n \rvert$; γ_n	The $\mathcal{S}$-cell set (together with its geometric realization and canonical s-block bundle) serving as universal object for n-dimensional s-block bundles.
1.13	$\tau: T^* \to Q_n$	Gauss map classifying the tangent s-block bundle of a locally-ordered triangulated n-manifold.
1.17	$qA(n)$; $\tau: T^* \to qA(n)$	classifying space, analogous to Q_n, for A-homology s-block bundles; Gauss map for A-homology n-manifolds with locally-ordered triangulation.
1.23	$\Theta(n)$	CW complex of n-dimensional Brouwer stars.
1.23	$B(n)$	simplicial space of Euclidean configurations of Brouwer stars naturally mapping to $\Theta(n)$.
1.23	$B(M)$; $g_M: M^n \to \to \Theta(n)$	If M^n is a locally-ordered Brouwer triangulated manifold, $B(M)$ is a construction analogous to $B(n)$ over $\Theta(n)$; the Gauss map $g_M: M^n \to \Theta_n$ is covered by $B(M) \to B(n)$.
1.24	N_K	space of normal planes to the general-position embedded Brouwer star K.
2.1	formal link (of dimension $(n,k;j)$)	represents possible local geometry of a simplex-wise convex-linear immersion of an n-manifold in R^{n+k} near an $(n-j)$ simplex.
2.4	$\mathcal{G}_{n,k}$	PL Grassmannian, with one j-cell for every formal link of dimension $(n,k;j)$.
2.6	$g: M \to \mathcal{G}_{n,k}$	canonical Gauss map from an n-manifold M immersed in R^{n+k}.

PAGE		REMARK
2.7-2.9	G: $\mathcal{G}_{n,k} \to R^{n+k}$	tautological map of $\mathcal{G}_{n,k}$ to R^{n+k}.
2.9	$\gamma_{n,k}$	canonical PL n-bundle over $\mathcal{G}_{n,k}$.
3.1-3.3	$\tilde{\mathcal{G}}_{n,k}$; $\tilde{\gamma}_{n,k}$	Grassmannian for simplicial immersions of triangulated n-manifolds in triangulated R^{n+k}, together with canonical k-block bundle.
3.5-3.6	$A\mathcal{G}_{n,k}$ ($A\tilde{\mathcal{G}}_{n,k}$)	Grassmannian for simplex-wise convex-linear immersions (simplicial immersions) of A-homology n-manifolds in R^{n+k} (triangulated). $A\mathcal{G}_{n,k}$ admits a canonical A-homology block-bundle $A\tilde{\gamma}_{n,k}$.
3.8	$\mathcal{G}^s_{n,k}$, $\gamma^s_{n,k}$	Grassmannian for arbitrary simplex-wise convex-linear maps from triangulated n-manifolds to R^{n+k}, together with its canonical PL bundle.
4.1	geometric subcomplex of $\mathcal{G}_{n,k}$	
6.2-6.3	$\mathcal{G}_{n,k}(W)$	Grassmannian "bundle" for immersions of n-manifolds into the triangulated (n+k)-manifold W; analogous to the Grassmannian bundle $G_{n,k}(W)$ of a smooth (n+k)-manifold.
6.5	G_W	tautological map from $\mathcal{G}_{n,k}(W)$ into W.
6.5	$\gamma_{n,k}(W)$	canonical PL n-bundle over $\mathcal{G}_{n,k}(W)$.
6.6-6.7	$g(f): M^n \to \mathcal{G}_{n,k}(W)$	Gauss map corresponding to an immersion $f: M^n \to W^{n+k}$.
7.2-7.3	$\mathcal{G}^c_{n,k}$	Grassmannian for piecewise-smooth immersions of triangulated manifolds and more general stratified manifolds.
7.3-7.4	$\gamma^c_{n,k}$	canonical PL n-bundle over $\mathcal{G}^c_{n,k}$.
7.5	$\mathcal{L}_T$	Space of formal links with a given combinatorics T; $\mathcal{L}_T$ is a subspace of $\mathcal{G}^c_{n,k}$ with a natural neighborhood N_T, itself the image of a disk bundle P_T.

A.3

PAGE		REMARK
7.8-7.11	strict stratification; stratification; linkwise-simplicial stratification	
7.13	$\mathcal{M} = \{M(X)\}$	codimension-0 decomposition of M corresponding to a stratifica-by strata $\{X\}$.
7.18	B_{LS}	Classifying space for linkwise-simplicial stratified structures on manifolds.
7.29	Gauss map of piecewise-differentiable immersion	map $M^n \to \mathcal{G}^c_{n,k}$ defined for a class of immersions satisfying α, β, γ of page 7.29.
7.32	Geometric subspace of $\mathcal{G}^c_{n.k}$	
7.34	$\mathcal{G}^c$	double limit $\lim\limits_{n,k\to\infty} \mathcal{G}^c_{n,k}$, analogous to BO.
7.35	Θ_i	group of equivalence classes of LS stratified i-spheres.
8.1	A^{ord}	classifying space for locally-ordered triangulated manifolds.
8.2	A^{Br}	complex of Brouwer structures on ordered triangulated spheres: [NB: A^{Br} is similar to $\Theta(n)$ of §1].
8.3	A^{cBr}	retopologization of A^{Br}.
8.4	$\mathcal{G}^{st}_{n,k}$; $\mathcal{G}^{cst}_{n,k}$	subcomplex of $\mathcal{G}_{n,k}$ arising from "straight" formal links; together with its retopologization as a subspace of $\mathcal{G}^c_{n,k}$.
8.5	$\mathcal{G}^{ord}_{n,k}$; $\mathcal{G}^{cord}_{n,k}$	slight generalization of $\mathcal{G}_{n,k}$ to take account of local orderings; together with retopologization analogous to $\mathcal{G}^c_{n,k}$.
8.6	$\mathcal{G}^{os}_{n,k}$; $\mathcal{G}^{cos}_{n,k}$	subcomplex, arising from "straight" links, of $\mathcal{G}^{ord}_{n,k}$; together with its retopologization [Note: $\mathcal{G}^{cst}_{n,k}$, $\mathcal{G}^{cord}_{n,k}$, $\mathcal{G}^{cos}_{n,k}$ are used in the proof of Theorem 8.1 but are not themselves of primary interest.]

PAGE		REMARK
10.1-10.2	$\mathcal{G}^c_{n,k}(W)$; $\gamma^c_{n,k}(W)$	$\mathcal{G}^c_{n,k}$ - bundle over a smooth Riemannian manifold W, together with its associated canonical PL n-bundle. $\mathcal{G}^c_{n,k}(W)$ is the target for the Gauss map of a piecewise differentiable immersion into W of an LS stratified n-manifold.

REFERENCES

[A-P-S] Atiyah, M.F., Patodi, V., and Singer, I., Spectral Assymmetry and Riemannian Geometry, Bull London, Math. Soc. 5 (1973), 229-234.

[Bi] Bierstone, E., Equivariant Gromov theory, Topology 13 (1974), 327-345.

[Br] Bredon, G., An Introduction to Compact Transformation Groups, Academic Press 1972.

[Bro] Brown, E., Cohomology theories, Ann. of Math. 75 (1962), 467-484.

[C_1] Cairns, S., Homeomorphisms between topological manifolds and analytic manifolds, Ann. of Math. 41 (1940), 796-808.

[C_2] ________, Isotopic deformations of geodesic complexes on the 2-sphere and plane, Ann. of Math. 45 (1944), 207-217.

[C_3] ________, Introduction of a Riemannian geometry on a triangulable 4-manifold, Ann. of Math. 45 (1944), 218-219.

[Ch_1] Cheeger, J., Topology of manifolds, (J.C. Cantrell and C.H. Edwards Jr., eds.) Markham, Chicago 1970, 470-471.

[Ch_2] ________, Spectral geometry of singular Riemannian spaces, J. Differential Geometry 18 (1983), 575-657.

[Gab] Gabrielov, A.M., Combinatorial formulas for Pontrjagin classes and GL-invariant chains. Functional Analysis and its Applications 12 (1978) 75-80.

[G-G-L] Gabrielov, A.M., Gel'fand, I.M. and Losik, M.V., A local combinatorial formula for the first class of Pontrjagin Functional Analysis and its Applications 10 (1976), 12-15.

[H-P] Haefliger, A. and Poenaru, V., La classification des immersions combinatoires, Inst. Hautes. Etudes. Sci., Publ. Math. 23 (1964), 65-91.

[H-T] Halperin, S., and Toledo, D., Stiefel-Whitney homology classes, Ann. of Math. (1972), 511-525.

[He] Henderson, D., The space of simplexwise geometric homeomorphisms of the 2-sphere, preprint.

[Hi] Hirsch, M., Immersions of manifolds, Trans AMS 63 (1959), 242-276.

[Hi-M] Hirsch, M., and Mazur, B., Smoothings of Piecewise-linear manifolds, Annals Study 80, Princeton University Press, Princeton (1974).

[La-R] Lashof, R., and Rothenberg, M., G-smoothing theory, Algebraic and Geometric Topology, Proc. Symposium Pure Math. XXXII, 1978 Part I, 211-266.

(cont.) REFERENCES

[Le_1] Levitt, N., the PL Grassmannian and PL curvature Trans. AMS 248 (1979), 191-205.

[Le_2] ________, Some remarks on local formulae for Pl, Algebraic and Geometric Topology, Proceedings Rutgers 1983, Lecture Notes in Mathematics 1126, Springer, Berlin, 1984, 96-126.

[Le-R] Levitt,N. and Rourke, C., The existence of combinatorial formulae for characteristic classes, Trans. Amer. Math. Soc. 239 (1978), 391-397.

[Mac] Mac Pherson, R., The combinatorial formulae of Gabrielov, Gel'fand and Losik for the first Pontrjagin class, Seminaire Bourbaki, 29, No. 497 (1976-77).

[M-M] Martin, N., and Maunder, C.R.F., Homology cobordism bundles, Topology 10 (1971), 93-110.

[Mi] Milan, L., Ph.D Thesis, University of Georgia, 1987.

[Mi-St] Milnor, J., and Stasheff, J.D., Characteristic Classes, Princeton University Press, Princeton 1974.

[Ml] Mladineo, R.H., Ph.D Thesis, Rutgers, 1979.

[P] Poenaru, V., Homotopy theory and differentiable singularities, Lecture Notes in Mathematics 197, Springer, Berlin, 1971, 106-132.

[R-S] Rourke, C.P., and Sanderson, B.J., Block bundles I,II,III, Ann. of Math. (2) 87 (1968), 1-28, 256-278, 431-483.

[St_1] Stone, D., Notes on "A combinatorial formula for $p_1(X)$", Advances in Math 32 (1979), 36-97.

[St_2] ________, On the combinatorial Gauss ways for C^1-submanifolds of Euclidean space, Topology 20 (1981), 247-272.

[St_3] ________, Sectional curvature in piecewise-linear manifolds, Bull. Amer. Math. Soc. 79 (1973), 4060-1063.

[St_4] ________,Geodesics in piecewise-linear manifolds, Trans. Amer. Math. Soc. 215 (1976), 1-44.

[T] Teleman, N., Global analysis on PL manifolds, Trans. Amer. Math. Soc. 256 (1979), 49-88.

[Wa] Wall, C.T.C., Classification problems in differential topology IV; Thickenings, Topology 5 (1966), 73-93.

[Whd] Whitehead, J.H.C., Manifolds with transverse fields in Euclidean Space, Ann. of Math. 73 (1961), 154-212.

[Whn] Whitney, H., On the theory of sphere bundles, Proc. Nat. Acad. Sci. USA 26 (1940), 148-153.